The Owl Who Liked
Sitting on Caesar

The Owl Who Liked Sitting on Caesar

Living with a Tawny Owl

MARTIN WINDROW

ILLUSTRATIONS BY CHRISTA HOOK

FARRAR, STRAUS AND GIROUX

NEW YORK

Farrar, Straus and Giroux
18 West 18th Street, New York 10011

Library of Congress Cataloging-in-Publication Data
Windrow, Martin.
    The owl who liked sitting on Caesar : living with a tawny owl / Martin
Windrow ; illustrations by Christa Hook. — First edition.
        pages   cm
    Includes bibliographical references.
    ISBN 978-0-374-22846-0 (hardcover) — ISBN 978-0-374-71153-5 (ebook)
    1. Tawny owl—Great Britain.   2. Owls as pets—Great Britain.
I. Title.   II. Title: Living with a tawny owl.

QL696.S83 W56 2014
598.9'7—dc23

                                                        2013036542

Designed by Abby Kagan

Farrar, Straus and Giroux books may be purchased for educational, business,
or promotional use. For information on bulk purchases, please contact the
Macmillan Corporate and Premium Sales Department at 1-800-221-7945,
extension 5442, or write to specialmarkets@macmillan.com.

www.fsgbooks.com
www.twitter.com/fsgbooks • www.facebook.com/fsgbooks

1   3   5   7   9   10   8   6   4   2

CONTENTS

A NOTE ON OWLS

READERS SHOULD BE aware that in the UK all birds of prey, their eggs, and their hatchlings are fully protected by law. If you chance upon what you think is a "lost" owlet in the spring, do *not* be tempted to "rescue" it and take it home with you. Only interfere if it is obviously in harm's way—on the ground, within reach of dogs and other predators. In that case, pick it up by cupping it softly in your hands, lift it onto a safe branch, and leave it for its parents to find (they certainly won't be far away)—or for it to climb back to the nest by itself, which it will usually manage perfectly well. It is a myth to imagine that owls will abandon their young if they become "tainted with human scent" by being touched. Traditionally, it has been believed that owls have almost no sense of smell; whether or not that's true, if they find their strayed fledgeling within about twenty-four hours the parents will continue to feed it.

Only if the owlet is clearly injured should you even consider taking it home. In such cases it is important to contact some qualified person—a vet, a local RSPB or RSPCA officer, but preferably a specialist bird rescue centre—*without delay.* Keep it in a good-sized cardboard box, open at the top.

If it is necessary to feed an owlet before somebody better

qualified can take over its care, do *not* feed it bread and milk, which will kill it; owls are exclusively carnivores, whose digestion depends on making use of all parts of their animal prey. If you have to feed an owlet, then offer it scraps of raw minced beef (repeat: beef—not all meats are safe), perhaps dipped in egg-yolk; use something like a blunt matchstick, and position the food well back in its mouth. A short-term solution to the important need for roughage is to mix in bits of soft feather (obviously, natural materials only—nothing that might have been dyed or otherwise chemically treated). But an owl's dietary needs are more complicated than this; enlist the help of an expert in the care of raptors, and get real-time advice—quickly. The following address and websites may be helpful:

Hawk and Owl Trust
PO Box 400, Bishops Lydeard
Taunton TA4 3WH
Tel: 0844 984 2824
enquiries@hawkandowl.org.uk

www.animalrescuers.co.uk/html/owls.html
www.barnowltrust.org.uk
www.raptorfoundation.org.uk/finding.htm

The Owl Who Liked
Sitting on Caesar

Introduction

SHAVING IS TRICKY with an owl on your right shoulder.

When I am working on the right side of my throat, Mumble tends to make darting, snake-like passes with her beak at the handle of the razor as it reaches the top of each stroke. Often, with a curiosity that disappointing experience never seems to dull, she takes opportunities, while I am working on the left side of my neck, to peck thoughtful gobs of shaving soap off the right side. The taste does not seem to appeal; after a few ruminative smacks of her mandibles she gives a little sneeze (*snit!*), and most of it ends up distributed around her whiskers. Still, she sometimes hops down to the edge of the basin and watches the floating clots of soap and bristles with interest. She feels delightful against my bare belly, warm and velvet-soft; but—given her lively and usually tactile curiosity—I get a little nervous when she heads that far south.

I have tried persuading her to stroll across behind my neck to the left shoulder when operations on that side are complete, but she's a right-shoulder owl by preference, and—like me—she does not welcome any kind of novelty at this hour of the day. We are both operating on auto-pilot, and this limited ability to cope with mornings is a bond between us.

The shaving mirror reflects two pairs of eyes—one pair bloodshot blue, one glassy black—side by side in a rather sordid mess of wet hair, soap, and feathers. I imagine that I can recognize in both pairs the familiar morning combination of apathy tinged with vague suspicion about what the day may bring: for me, sinister buff envelopes with cellophane windows; for her, perhaps, a troublesome frayed feather among the left wing secondaries. Who am I to add to her problems by trying to force her to cope with a radical new concept like helping me to shave from the left shoulder? We manage; in fact, we manage so well that usually I don't even notice the bizarre adjustments into which I have gradually slipped during the three years since we first met.

OCTOBER 2013

MUMBLE WAS SO much a part of my life in those days that the oddity of our relationship seldom occurred to me, and I only thought about it when faced with other people's astonishment. When new acquaintances learned that they were talking to a book editor who shared a seventh-floor flat in a South London tower block with a Tawny Owl, some tended to edge away, rather thoughtfully. Collectors of eccentrics were pleased—some of them to the extent that for years afterwards I was deluged at Christmas and birthdays with owl-related greetings cards. (This was mildly endearing at first, but got rather wearying in the longer term.) However, more conventional-minded people might question me—

sometimes, I thought, rather relentlessly—about the practicalities of my domestic arrangements. I tried to answer patiently, but I found it hard to come up with a short reply to the direct question "Yes, but . . . *why?*"; my best answer was simply "Why not?"

I am embarrassed to recall that on one occasion I tried an unattractively smart-arse routine: "Look—I've lived with her for two years. She costs me about twenty quid a year, all in. She's wonderfully pretty, and amusing. She's affectionate without being needy, and she smells great. She doesn't mind how late I get home, she doesn't talk at breakfast time, and we hardly ever argue over who gets which bit of the Sunday paper." When I thought over what this rant might suggest about my attitude towards human female company, I swiftly dropped it from my conversational repertoire.

Almost invariably, when people actually met Mumble I didn't have to say anything more to convince them. Whatever their preconceptions, when they were first confronted by a Tawny Owl at close range, their faces would instantly light up and soften. During her first year or so, when it was still possible for her to meet strangers without glass or wire mesh between them, then—unless I remembered to caution them—their first wondering exclamation (usually along the lines of "Oh! . . . but it's *gorgeous!*") would often be accompanied by an instinctive reaching out to stroke her.

Rather less welcome was my discovery that if I met that person again only after an interval of years, the first thing they tended to say was "Oh yes, of course—the owl man!" I have since consoled myself with the thought that there are far worse reasons for being (however slightly) memorable.

◎◎

THE STRICTURES IN the Note on Owls against giving way to a temptation to "rescue a lost owlet" might seem hypocritical in a book about the pleasures of living with an owl, but my justification is that Mumble was not taken from the wild. She was hatched in captivity and reared by hand, and never knew a relationship with her own kind. I was able to give her a much longer, better-fed, and less dangerous life than she would have known in the woods. At first I did feel an occasional twinge of guilt about denying her "the freedom of the skies," but I soon found out that in the case of a Tawny Owl such feelings have everything to do with human sentimentality and nothing whatever to do with Nature—a tawny isn't a skylark or a peregrine falcon; it's a home-loving cat with wings. On the couple of occasions when she had the opportunity to do so, Mumble failed to show the slightest interest in exploring the freedom of the skies (and in the end, it was probably somebody in the grip of that sentimental delusion who caused her premature death).

My feelings after that event were one of the reasons why, despite periodic nagging from my family, it has taken me many years to finally get round to digging out the notes and photographs that I took during the fifteen years that Mumble and I spent together, and to try to turn them into this book. I was a younger man, and so an angrier one. I convinced myself, on fairly scant evidence, that the intruder who had cost my owl another decade or so of contented life must have been some "animal rights" activist, more self-righteous than well-informed. (For several decades now, huge numbers of the smaller animals and low-nesting birds in England have been massacred by an alien predator—the mink, released by activists who raided fur farms. Their motives may sometimes have been compassionate, but their arrogant disregard for the law of unintended consequences is hard to forgive.) Anything I had tried to write about Mumble then would probably have been distorted by my rage over what I then imagined were the circumstances of her death.

Time brings calmer judgement, and when I found myself with a gap between writing projects connected with my profession, it occurred to me that this might be the time to re-explore the "owl years." Since I finally began to reread the notebooks that I had laid aside in the mid 1990s, I have found myself reliving emotions that I had long locked away—and I am glad that I have.

I should make one point about the text that has emerged from this process. I don't pretend that all the "diary" entries at some places in this book are taken literally verbatim from my notes made at the time, although I did work up many of those at some length when I first wrote them. I have naturally edited and elided some others; but all of them are faithful quotations from what I jotted down shortly after the events or thoughts that they record.

◉◉

WHY I DECIDED, in my thirties, to acquire a pet for the first time—and an owl, at that, given my previous complete lack of interest in ornithology—remains a fair question. If the reasons why were puzzling, then the how was also less than straightforward.

Mumble was not, in truth, my first owl; and although she became the epitome and icon of "owlness" for me, it would be dishonest to airbrush out of the record my failed first relationship. Like most such mistakes, it taught me a lot.

1

◎◎

Man Meets Owl—Man Loses Owl—
Man Meets His One True Owl

I T ALL BEGAN, as so many things have done over the past half-century, with my older brother, Dick.

By the mid 1970s he had achieved his long-held ambition to move out into the Kent countryside and acquire as old a property as he could, with enough space to indulge his several hobbies at weekends. (The list has included, over the years, competitive rally-driving, military vehicle restoration, aviation archaeology, rough shooting, and falconry, besides guitar blues and various other pastimes involving extraordinarily exact and fiddly work for a man with large paws.) Since his wife, Avril, was both patient and highly competent in a wide range of practical skills (from fine needlework and working in silver, through gardening and animal husbandry, to concrete-mixing, structural repairs, and decorating), Water Farm soon became a very attractive and interesting place to spend time, even though the farmhouse's immediately previous occupants had been sheep. Moreover, there were very few commodities or services that you could discuss with Dick without seeing a thoughtful look steal over his friendly, slightly battered countenance: "Ah, now that's interesting— as it happens, I know this bloke who . . ." (can supply an army-surplus tank engine, cures sheepskins, works as a film stuntman, knows when his lordship's warrens will be unguarded for the weekend, understands explosives, breeds wild boars, speaks Dutch, casts things in fibreglass, can get you whatever-it-is without the bother of boring paperwork, etc., etc.).

At that time I was living in a high-rise flat in Croydon, South London, and commuting daily to a publisher's office in Covent Garden, where I worked as the commissioning and art editor of a military history book list. In those days the extended family usually spent Christmases at Water Farm, and—since I both lived and worked surrounded by dirty concrete and diesel fumes—I would often exploit Dick and Avril's endless hospitality to spend summer week-ends in the Kent countryside. They kept a variety of animals over the years, off and on: multiple cats (including one who proved humiliatingly better than me at hunting rabbits), doves, chickens, ducks, geese, turkeys, a few sheep, a goat, a donkey, a Dexter-Angus cross cow, my nephew Stephen's splendid polecat-ferret Shreds, and for a time even a raccoon (fully grown, they're a great deal bigger and stronger than you might think). I was not particularly an "animal person," but this menagerie certainly added to the attractions of peace, space, clean air, and Avril's magnificent cooking.

Before they even moved to Water Farm, Dick had become interested in books on falconry. Inevitably, he soon made friends in that world too, and acquired his first bird—a sleekly beautiful Lanner Falcon named Temudjin, after the young Genghis Khan. After buying the farm he built a mews and flights (living quarters for falcons, and aviaries large enough for them to move around in), and as his knowledge, circle of acquaintance, and skills all increased, these quarters came to be occupied by a succession of hunting birds. They included kestrels, buzzards, goshawks, and even a second-hand Steppe Eagle suffering from something called "bumble-foot" (no, me neither).

Watching Dick handle and train these lovely creatures,

it was impossible not to become intrigued myself. When I was finally allowed to pull on a glove and take one of them for a supervised stroll through the fields and lanes, the medieval spell brushed me at once. It's a hard feeling to describe. There was vanity, of course: I refuse to believe the man lives who would not find himself striking a Plantagenet pose and unconcernedly stroking his falcon's breast when a turn in the lane reveals a couple of gratifyingly impressed young female ramblers. But it was more than simply ego; it was a new kind of relationship for me, which put me in touch with a different set of feelings. I suspected that they went deep, and came from somewhere very old. It was a slow process, which I did not even admit to myself for some months, but gradually I began to realize that I wanted some sort of lasting contact with this new thing.

The idea of keeping a falcon in a high-rise apartment block in South London was obviously ridiculous, but the daydream would not let me alone. It was my sister-in-law who unwittingly showed me the way in. Avril had wanted a bird of her own for some time, but one that could be fitted into the routine of the tirelessly active mother of two boys. Dick duly made a number of phone calls to gentlemen with odd nicknames, and in the fullness of time "Wol" took up residence in Avril's kitchen, spending most of his time on a perch in the shadows on top of a tall cupboard. Avril's kitchen was a welcoming haven for casual passers-by, and the addition of a Tawny Owl simply added to its attractions. (Wol sat so still that he was always assumed to be stuffed, until an eventual blink gave the game away; this had been known to cause a visitor to spill coffee or choke on a mouthful of cake.)

I was charmed by Wol from the moment I laid eyes on him, and as it became plain how easily and unhysterically an owl—if taken young enough—can grow accustomed to human company, my resistance to the nagging idea of getting a bird for myself weakened.

◎◎

IN THE SUMMER of 1976 a friend and I begged spare beds at Water Farm while we attended a short parachuting course at a nearby airfield.

This was long before novice sports parachutists had access to modern rigs with their relatively light packs, mattress-shaped canopies, and sensitive controls that allow you to make a "stand-up" landing almost every time. Roger and I were taught how to make the landing-rolls that were necessary with the old World War II–vintage Irvin chutes with the X-type harness, which seemed to weigh as much as (and brought you to earth with all the cat-like grace of) a sack of potatoes.

My first jump was terrifying and exhilarating in equal measure. First came sheer, abject, bowel-loosening terror, as the engine of the little Cessna was switched off and I had to clamber out and balance between the wing strut and the landing gear, straining to make out the jumpmaster's reminders above the rushing of the wind. Then—when the canopy had slammed open, the tight harness was holding me like the hand of God, and Kent was smiling up between my feet—came a wave of sheer exhilaration, which redoubled when I struggled to my feet after managing a successful landing.

However, it was the third jump that proved to be the most memorable experience. With the positively eerie lack of physical co-ordination that had been so noted by sports masters during my schooldays, when the final "green rush" snatched me down, I misjudged my roll spectacularly. I hit the ground backside-first, thus guaranteeing one of the classic (and excruciatingly painful) parachuting injuries—a compression fracture of the lumbar vertebrae. The luckless Roger, who had drawn second straw and was still some hundreds of feet above the drop zone, had to prepare himself for his own landing while being distracted by my noisy writhings. My most lasting memory of the next half-hour is of a young pre-jump army cadet among the circle of anxious figures staring down at me. He put a cigarette in his mouth, patted his smock pockets distractedly, muttered to his comrades—who shook their heads, unable to take their solemn eyes off me—and then leaned down to ask me if I had a light, mate. Temporarily preoccupied with thoughts about my spine, I was unable to oblige him.

In June 1976 southern England was sweltering under a once-in-twenty-years heatwave, and the hospital bed in which I was trapped, sweating freely and unable to move an inch, was immediately below a large skylight in the low ceiling of a single-storey side ward. Staked out under the burning sun like a victim of the Apaches, and unable to face the truly hideous hospital slops, I coped partly thanks to a kindly veteran night nurse with a relaxed attitude towards injecting Demerol, and partly thanks to Dick, who faithfully visited me on his way home from work every evening, bringing delicious sandwiches. After a week, corseted in sweaty canvas and metal splints, I was able to lurch slowly out to his car, moving like Boris Karloff in *Frankenstein*, and was carried back to Water Farm to convalesce.

◎◎

UNAVOIDABLY, OVER THE weeks that followed I spent many hours lying on a blanket in the shade with a book, or tottering slowly around while I recovered my mobility. I had more time to watch Dick's birds than ever before, and my interest grew. Even I couldn't spend entire days reading books without a break, and the birds became a welcome distraction. With the time to simply be still and observe them, and to revisit them for fairly long periods several times a day, I began to get a sense of the rhythm of their lives rather than just a series of snapshots. Watching them preening themselves brought the detailed structure of their bodies into closer focus, and I started to notice their individual characteristics. I began badgering my brother with questions about their quarters, food, daily routines, medical and emotional needs, and other anticipated requirements, some of them no doubt extremely silly.

These conversations continued intermittently by telephone long after I had returned home. If Dick had agreed with my often expressed doubts about the whole idea, I would probably have given up; but he isn't the kind of person who assumes in advance that any dream, however daft, is unattainable. Before long I was running out of arguments against myself, and the evening came when I took a deep breath and asked Dick to telephone "this bloke he knew." Perhaps with some vague idea that if keeping an owl proved to be a disaster, then a small owl would make it a small disaster, I asked him to find me a Little Owl (this is a species, not a description).

And so it was that in the autumn of 1977, a six-inch, four-ounce bundle of feathered fury took up residence with me on the seventh floor of a large concrete apartment block beside the A23 in West Croydon. With his distinctly hawkish profile, beetling brows, and blazing yellow eyes, his name could only be "Wellington." Unfortunately, he also turned out to share the Iron Duke's stubborn willpower.

◎◎

THE THRUSH-SIZED LITTLE OWL—*Athene noctua*—is the smallest of Britain's owls, and the most recently arrived. It was introduced during the second half of the nineteenth century from continental Europe, by landowners attracted by its reputation as the scourge of mice and of insect pests; in several European countries Little Owls are actively encouraged by farmers, and protected by law. There is an attractive story that the first Englishman to exploit their usefulness was Admiral Nelson. When he was serving in the Mediterranean he is supposed to have acquired one hundred Little Owls from North Africa and given them to each of his ships; they were kept on the officers' tables at mealtimes, to clear the weevils out of their spoiled ship's biscuit. (I have no idea if this tale is true, but I would love to believe it. I can just *hear* Nelson's seadogs cheering on their owls and laying wagers on how many prizes each would take.)

The current British population is estimated—with the usual airy lack of precision found in all such figures—at anything between five thousand and twelve thousand breeding pairs. It has diminished over the past few decades, and is now on the conservationists' "amber list" as a species causing "moderate concern." They are the least nocturnal of our owls, and although they do hunt after dark they are also active during daytime. Little Owls have barred and mottled dark brown and white plumage, and a rather more streamlined silhouette than larger species, with a flatter-looking top to their heads. They have the broad, rounded wings of a woodland bird, and a very short tail. In Europe they prefer to live in woods and copses among farmland, and as you drive through the English lowlands you may occasionally spot a hunched little figure sitting on a fencepost, checking out the open terrain of fields and hedgerows. At the right times of year they may even be seen following the plough to catch worms.

The first of my many mistakes had been in asking for this breed of owl at all, and worse still was the fact that this particular owl was already six months old, and had spent those months in a large aviary with other birds. The most basic rule when seeking to tame any wild creature is that it should be isolated from its kind and raised by the handler from the earliest possible age—as soon as it can safely be separated from its mother. With careful kindness, the animal may be persuaded to project onto the handler any potential it has for social feelings. It is widely understood that a truly social animal, such as a dog, can easily be trained to regard its human owner as the "alpha dog of the pack." A solitary hunting bird—a raptor, such as an owl—feels no such instinctive connection. The egg must be taken from the nest and hatched in an incubator, so that as soon as it emerges from the shell the hatchling sees, and is fed by, a human.

It has sometimes been said that the bird will then "imprint" on this person, forming an unbreakable bond and making it impossible to return it to the wild. That is to overstate the case by a wide margin. A fledgeling raised for its first few weeks by one handler can easily transfer this familiarity to another human. Foundlings raised by humans have often been reintroduced to the wild successfully by a process of gradual disengagement. Alternatively, if carefully introduced to an aviary with other birds, then in time they become accustomed to the company of their own kind. However, if the bird spends the formative first weeks of its life outside the egg among other birds, and without being handled by a human, it is widely believed to be more or less untamable. This was the case with Wellington; I ought to have known that my attempts to "man" him—to tame him to my touch—were probably doomed from the start.

◎◎

BECAUSE WELLINGTON WAS a nervous wild creature, unused to being handled, he had to be "jessed" like a falcon before I took him home with me, or he would have been impossible to control.

Jesses are narrow strips of thin, light leather that a falconer fastens round his bird's ankles so that it can be held by them as it sits on his fist. The trailing ends are united with a little metal swivel-ring device (for falconry, a pair of tiny brass bells are also attached). When the falconer passes a leash through the swivel-ring—a yard or so of cord, with a stop-knot at the bird end—he can tether it to another swivel-ring on its perch or on a "weathering block" in the open air. This leaves the bird with plenty of room to move around but no chance of tangling itself up in the leash (or that's the theory, at least; in practice, some birds seem able to defeat this supposedly foolproof design with laughable ease).

Fitting jesses to an untamed bird is obviously a job for two pairs of hands—in the case of Wellington, belonging to one expert and one apprehensive novice. He had to be taken out of his cage and held passive, lying on his back, with his legs in the air and his wings held gently but firmly to his sides—if he could get a wing free and start lashing about, we were in trouble. Some people favour holding birds swaddled in a soft cloth, while others are confident enough to take the correct hold with bare hands. As a nervous apprentice, I found it a disquieting job: until I had done it a few times I didn't have an instinctive feel for how and where to grip. I was naturally terrified of holding too tightly—any constriction of a bird's chest can be fatally dangerous—and I was taken aback to discover just how strong and wriggly such a little bird could be.

If you get it right the bird just lies there, perfectly safe and comfortable, but a picture of outraged dignity. Personally, I always felt embarrassed and apologetic at that point, but this obscure sense of moral inferiority to the bird can disappear in a hurry if it manages to get a foot into you. Even the smallest raptors have extraordinarily powerful talons, and if they connect, they hurt. One of the tricks Dick taught me was that a bird that's feeling spiteful can be given a pencil to hold: as soon as this touches its feet the wicked hooks snap closed around it and hang on to it like grim death while you get on with fitting the jesses. (Since these inevitably get worn and tatty with badly aimed droppings, the remains of food, and frequent absent-minded chewing, the chore of fitting your bird with nice clean socks has to be repeated at fairly regular intervals. However tame and lazy you think it has become, it can still give you a painful surprise if you relax your concentration during this procedure.)

◎◎

I DROVE BACK to London that first Sunday night with Wellington on the passenger seat in a fairly large cage provided by Dick. Carrying it up from the underground parking garage into the block of flats, and up in the lift to my floor, took several nerve-racking minutes—one of my more rational misgivings about this whole project had been that all pets were actually forbidden in these flats. The caretaker was an uncompromising Yorkshireman who ran a tight ship, and one or two minor incidents in the past, when I had been sharing the flat with a journalist former workmate of mine, had led him to regard flat 40 with a distinctly jaundiced eye. (In our defence, I must protest that Roy and I had seldom thrown parties—but when we did, we prided ourselves on giving our guests a good time.)

Luckily, on that evening the lift passed the caretaker's floor without the stop-button lighting up. Once safely indoors, I put the cage on a table in the living-room, where Wellington was going to live until I built him something more spacious. I put some straw and newspaper inside for reasons of hygiene, and gave him a split log to sit on. The cage was a wide wooden box with a wire-mesh front, so he had a good field of view while still feeling the security of a roof and walls around him. This seemed sensible; in the wild, *Athene noctua* nests in tree holes, odd corners of farm buildings, or even down abandoned rabbit burrows.

On the first few evenings when I got home from work I would be greeted by Wellington's fierce yellow eyes blazing defiance from the shadows behind the wire mesh. After eating supper myself I would get some food out of the fridge for him, and settle down for the first sessions of trying to "man" him. In the wild Wellington's diet would have consisted largely of insects, though he would have relished anything from craneflies, earwigs, beetles, moths, worms, slugs, and snails, up to small rodents. Having been raised in captivity, however, he was accustomed to the usual rations for captive birds of prey: dead day-old chicks, which are handy little packages of nutrition still with egg-yolk in their body cavities. Chicken hatcheries always have large supplies of these unwanted male chicks, and have learned that they can make a few pounds by refrigerating sacks of them for sale to falconers. Dick had given me a couple of dozen to keep Wellington going until I found a regular supplier of my own through the Yellow Pages.

◎◎

THERE IS SELDOM any secret to taming a wild animal, beyond common sense and kindness. You have to handle them gently and repeatedly until they lose their fear of you. You have to be endlessly calm and patient, because if you project fear or anger you can set the process back by days.

This is, of course, especially true of a solitary animal as opposed to a pack animal: while a puppy has the mental mechanism to understand the concept of "correction," and will make a submissive response, a hunting bird interprets any sudden move as simple aggression.

You have to use their hunger to entice them to tolerate you; hunger is at first your only way of creating any kind of transaction between you. "Hunger" means appetite—not starvation. Apart from being cruel, starvation is obviously counter-productive: you are trying to create a mood of calmness, and what starving creature is calm? Birds of prey consume a lot of fuel and so need careful feeding, and by learning to regulate the amount and timing of the daily meal it is usually possible to establish some kind of routine relationship fairly quickly. (I should emphasize here that I am talking about taming a bird as a pet, not the much more complex process of training it to hunt free. True falconry involves calibrated feeding and regular weighing, calculating the bird's rations to keep it healthy but "sharp set," so that it will be strong but still keen to hunt.)

◎◎

ALL I WAS hoping to achieve with Wellington was a basic level of tameness. I wanted him to learn to come to me of his own free will, at first for food and later, perhaps, simply to a call or whistle. I wanted him to lose his street-fighter wariness and allow himself to be played with and enjoyed. It seemed a reasonable target to set myself. After all, I had watched Dick make it seem ridiculously easy; he had once trained a kestrel indoors to come to his fist for food in less than a week, so I thought I knew roughly how to go about this game.

First spreading a newspaper on the floor by my chair and an old towel over the arm, to guard against accidents of a scatalogical nature, I would slip my left hand into an old driving glove (you don't really need a glove for protection with little birds like Wellington, but it gives them a better grip when they are standing on your fist). Then, with a boot-lace leash between my teeth, I would ease the cage door open a few inches and grope hopefully inside, trying to get hold of Wellington's dangling jesses and swivel, while he pranced and hissed his way around the most awkward corners of the cage. Finally getting a grip, I would gently pull him out until he gave it up as a bad job and jumped up onto my left fist. With my other hand I slipped the leash through the swivel-ring and wound its hanging end loosely round my fingers, holding the swivel firmly between thumb and forefinger until I was settled in my chair and could give him a bit more rope.

The purpose of the exercise was to accustom him to my company to the point where he would take food from my fingers and eat it on the glove. I hoped that when we had achieved this I could let him fly loose around those parts of the flat where he couldn't do much damage or injure himself, enticing him back to my fist with occasional treats. I would give a particular whistle whenever I showed him a snack—and only then—and I hoped that in time he would come to the whistle alone, bribe or no bribe. We would then be well on the way to the sort of relationship that I confidently expected.

The problem was that Wellington clearly hadn't been at-

tending when I explained all this. Night after night, week
after week, he would crouch (briefly) on my fist with all the
relaxed confidence of a scrap-metal dealer confronted by
auditors from Her Majesty's Revenue & Customs. For a
bird of Wellington's diminutive size I had to cut up the
squidgy, yolk-filled chicks with scissors—a repulsive task.
Suppressing my shudders, I would take a slimy gobbet from
the saucer by my side and hold it out to him, whistling and
crooning with what I hoped was seductive charm. Wellington would bob and weave, deliberately avoiding it and keeping his beak welded firmly shut, like a toddler watching the
approach of a spoonful of creamed spinach. I would dangle
the disgusting treat before his furious eyes; I would rub it
on his beak; after an hour or so I could barely resist the urge
to prise his stubborn mandibles apart and shove it in with
the end of a pencil. All was to no avail; unlike his glorious and
convivial namesake, Wellington dined alone in his quarters,
or not at all.

THE BORROWED BOX-CAGE was obviously only a temporary expedient. So that Wellington did not have to be confined whenever I was not holding him on his leash, I first
built him a "cadge." This was simply a portable table-top
perch, mounted in a tray wide enough to catch natural fall-
out and to allow short strolls on his leash. A seed box, a bit
of log fitted with a small swivelling ringbolt for the leash,
and last week's *Sunday Telegraph* were soon assembled and
set up on top of the box-cage. Tethered there during his first
weekend in the flat, he could watch further developments.

My plans for Wellington's permanent quarters were dictated by the layout of my flat. From the windowless, L-shaped entrance hallway the first pair of rooms—the bathroom and the bedroom that I used for my office—led off to left and right, the latter with a window overlooking a small balcony. Beyond these doors the hallway led on to my own bedroom straight ahead, with the kitchen to the left, and the large living-room to the right. This latter had been the main reason that Roy and I had chosen the flat (frankly, we had hoped it would be a girl-trap—and so it proved, though more often to Roy's benefit, dammit). The living-room was big, light, and airy, with an almost complete wall of floor-to-ceiling windows along the south side. This overlooked an open vista of tall buildings against a big sky—a sort of mini-Manhattan view, equally impressive in bright sunlight or spangled with lights after nightfall. The room caught the sun all day long; at the far end another wide window faced westwards over low roofscapes, towards the rising green swell of the old Croydon airfield a couple of miles away. (If the flat had existed in August 1940, it would have afforded a striking view of Hawker Hurricanes "scrambling" through the smoke from blazing factories to intercept German fighter-bombers.) To the right of this end window, a glass door led sideways out onto the balcony outside the office window. This was really only a glorified concrete shelf, darkly roofed in by the balcony of the flat above, but it was big enough for a couple of deck-chairs and a case of beer on sunny afternoons.

What I wanted to construct was a cage that would fit on the balcony, about the size of a large wardrobe and so big enough to allow Wellington to fly up and down for a few wing-beats. He could doze there in the fresh air while I was out at work during the day, and it would give him a more interesting view by night, all the while being sheltered from heavy weather by the overhang of the balcony above. This would put him within a couple of feet of the bedroom window of the flat next door, but luckily my neighbour Lynne was a good friend, who had no more love for the caretaker than I did. I assured her that owls of the species *Athene noctua* are not known for their loud singing at night. This claim was more hopeful than confident, but it turned out to be true. Since Wellington was so far from—and so far above—his natural farmland habitat, he had no real reason to issue his yelping territorial challenges, and there were no others within earshot to answer him.

◎◎

I COVERED SHEETS of graph paper with scribbles before a satisfactory blueprint emerged. Since the balcony was small and the door led out to it end-on, the planned creation could not be more than two feet wide if it was to leave space for me to squeeze out there past the nearside end of it, but there was room for it to be six feet long by six feet high. I planned a complete plywood section at the far end, about the size of a telephone kiosk, incorporating a hutch into which Wellington could retreat when he was feeling unsociable (which seemed to be his default setting), with a perch just outside his "doorstep" and a shelf for him to eat on. The rest of the structure was to be of wire mesh on a timber frame, with a couple more perches, made from branches, slanting across the corners at different heights.

I am emphatically not a handyman, but I thought that the door I devised was a stroke of genius almost worthy of an approach to the patent office. The Windrow Mark 1 Double-Reciprocating Owl Valve was made of mesh on wooden frames matching the inner dimensions of the cage, and was mounted towards the nearer end of its long "front" side. It was in fact two doors set back to back and hinged together, one opening inwards and one outwards. All I had to do was make sure that Wellington was down at the far end of the cage before pulling a wire to close the inner door across its width, shutting him down there. I could then pull open the outer door, enter, and close it behind me, thus shutting myself inside an "owl lock." There was just enough room in there for me to swing the inner door back past me again, leaving me and Wellington in the same space, without his ever having had less than one door between him and the open air. I would then get him into a basket, and reverse the process to carry him indoors with me.

Flushed with triumph, I set out on Saturday morning to the local do-it-yourself store. This had everything that I needed, but there were one or two aspects that I had not thought through—principally, the difficulty of manoeuvring eight six-foot lengths of one-inch by two-inch timber, three enormous sheets of marine-quality plywood (nothing but the best for Wellington), and sundry rolls of wire mesh through a narrow check-out exit, all balanced on a flimsy supermarket trolley. Act Two of this comedy of cruelty took place in the car-park, where I faced getting it all into or secured on top of my car ("Mummy, why does the funny red-faced man with the bleeding hands keep breaking bits of string and saying rude words?").

At noon, calmed by cool beer and the lack of a human audience, I laid everything out on the living-room floor and set to work. My plan was to make each side, each end, and the top separately, before juggling them out onto the balcony for final assembly. The scheme was sound enough, but the next thirty-six hours cruelly exposed my limitations as an entirely self-taught carpenter. Every measurement I had taken turned out to be that vital half-inch too short or too long. Every snipped end of wire mesh managed to stab my palm. Every staple bent uselessly as I tried to hammer it into the cheap, knotty wood. The hinge-screws split the timber, and at about 1:00 a.m. on Sunday morning I discovered that I was missing three of the necessary angle-brackets.

By Sunday evening I was dumb and savage with fatigue, and coated with sweat and sawdust, while the living-room floor looked like a building site—but there, by God, it stood at last. It glowed a mellow gold under its coats of weatherproof varnish, its Double-Reciprocating Owl Valve worked as if pivoting on silk and graphite, and its floor was lagged with snowy newspaper—a veritable owl-palace. Wellington was duly installed, the picture of haughty ingratitude. Hardly had I poured another beer and sat down before the doorbell rang. It was Attila the Caretaker.

He and I had been distinctly wary of one another ever since my moving-in party many years before. All too conscious that Wellington blatantly breached a clear term of my lease, I summoned up my courage and prepared to make the opening address for the defence. After what I had been through that weekend I was insanely prepared to resist the oppressors to the High Court and beyond, but—to my astonished relief—there was no word of illicit livestock. It

seemed that the trouble arose from another overlooked
schedule or sub-clause of the lease, which forbade me to
keep anything on the balcony that visibly lowered the tone
of the apartment block. Since my balcony overlooked a gas-
works, set among depressing terraced streets of two-up-
two-down houses that were due to be demolished within
the year, it was hard to imagine what I would have to do to
genuinely spoil the aesthetics of the area. Nevertheless,
the caretaker bluntly required me to move my "cupboard,"
sharpish. In my relief that he had not investigated further, I
promised to co-operate.

Early the next morning I slipped down for a street-level
reconnaissance. After lurking around several corners at
varying distances and angles from the block, I discovered
that the top part of the cage was indeed visible, but only at
ranges of 150 yards or more. It was almost entirely hidden
by the balcony parapet and the cave-like shadow cast by the
balcony above mine, and only the fine, glowing colour of
the kiosk at the end caught the eye. Sighing for my three
coats of carefully brushed marine varnish, I mooched off to
buy half a gallon of matte black paint. By lunchtime a second recce had reassured me that Wellington's quarters were
now completely invisible from ground level. Since the regime did allow plantings on the balconies, I also invested in
a fast-growing Russian vine and a bucket of compost. Within
weeks this provided Wellington with a second layer of camouflage, and I would hear no more from the landlord's
apparatchik.

◉◉

A COUPLE OF weeks later the problem of laying in rations for my owl became acute. I had found the number of a reasonably local hatchery, and had confirmed that they could sell me a sack of chicks, but I would have to turn up to collect them early in the morning before the pig-swill man arrived to take the lot. Since I was chasing deadlines at work too hard to take a morning off, that meant I could only fetch them on a Saturday. By that Tuesday I only had a couple of chicks left, and emergency measures were called for.

Presumably, I thought, Wellington could be tempted by some other sort of meat, so long as it was bloody, cut up small, and came complete with the fur or feather roughage that his digestive system demanded. What about raw rabbit—surely that was suitably rural? I remembered from my childhood that autumn was the time when one saw rows of shot rabbits hanging up outside butchers' shops, with tin cups under their muzzles to catch any dripping blood. In London the place for all kinds of specialist food suppliers servicing the restaurant trade was Soho, and that was only a ten-minute stroll from my office in Covent Garden. That lunchtime I walked over and began searching for an old-fashioned butcher selling game *au naturel*. The hunt was in vain; modern city taste apparently frowned upon such stark reminders of where meat actually comes from.

By the following day I was getting worried and decided to look further afield. It was a long-established legend that Harrods department store in Knightsbridge could supply anything that the heart desired, and the Harrods food hall was famous. I took a long lunch break and rode the Tube westwards. When I entered the hushed, high-ceilinged temple to gastronomy and found my way to the butchers' counters, sure enough, there was a row of rabbits hanging neatly against the marbled wall. They even seemed to have a well-mannered look, in keeping with their surroundings.

I had been so fixated on my target, and was so relieved to see it, that I had not actually thought out what I would say until I was approached by one of the staff. He was a silver-haired gentleman, impeccably dressed and aproned, with the quiet dignity of a bishop.

"And how can I help sir today?"

"I'd like a rabbit, please."

"Certainly, sir. Would sir be requiring a farmed or a wild rabbit?"

"Umm . . . what's the difference?"

(Very slightly pitying look:) "Generally, sir, it is believed that farmed rabbits are larger and more tender, but wild rabbits have more flavour."

"Er . . . a farmed one will be fine."

"Certainly, sir." The bishop turned and unhooked one of the furry corpses behind the counter. "And sir would like this skinned, no doubt?"

"Oh, umm—no, no thanks—leave the skin on, would you?"

"Of course, sir." He began to wrap the rabbit. But then I remembered, from occasional shooting forays from Water Farm, just what it took to actually dismember a rabbit from scratch—and I realized that I didn't have any knives in my own lazy bachelor's kitchen that would be strong enough for the job. This was going to be embarrassing.

"Umm—I say—er, could you possibly chop it up for me? With the fur on, I mean? Into fairly small chunks?"

The bishop stood quite still, his glossily shaven features

expressionless, and gave me a long look. Speaking rather slowly and deliberately, he asked:

"So sir would like me to cut up this rabbit—on the bone—unskinned—into *chunks* . . . ?"

"Yes—yes, would you, please? . . . Umm—it's not for me, you see." I was about to explain who it *was* for when my nerve failed me.

With his back to me, he proceeded to do as I asked. In long years of service to a demanding public, I imagine that this may have been a first for him. As his cleaver rose and fell, one of his fellow prelates was busy at the marble cutting-top next to him. I saw their heads move together. A moment later the other butcher stole a quick, inscrutable glance at me over his shoulder. The minutes seemed very long before I could thrust a banknote over the counter and make my self-conscious escape. It would be some months before I dared return.

◎◎

I NEVER DID find out if Wellington actually ate much of his painfully acquired rabbit. He certainly wouldn't touch his bunny-chunks during our next couple of frustrating evening sessions, so I had to leave bits of them in the balcony cage when I put him out for the night. He may have deigned to try them when he was alone, or he may have kicked them into a corner of his hutch with a hiss of contempt. That Saturday I bought several months' worth of dead chicks, and had to empty my small freezer to pack them away. (And yes—on one occasion a guest seeking ice-cubes for her gin and tonic did make a startling discovery.)

Wellington in his balcony cage, staring down at me with
his usual air of angry disdain.

The weeks of autumn turned into winter, and almost
every evening Wellington and I continued our battle of
wills as I persevered in trying to tame him. The sign I had
hoped for was the classic indication that a bird has decided
you are nice to be near: fluffing up his feathers, standing on
one leg, and taking a nap on your fist. Avril's Tawny Owl,
Wol, spent most of his life in this pose. Even Dick's falcons,
who were a right bunch of knife-fighters, would only sim-
per sleepily while he stroked their breast-feathers. But at my
gentlest touch Wellington went as rigid as a Victorian virgin,
and deliberately fell off my hand.

This manoeuvre is known in the trade as "bating off."
When a spooky bird takes it into its head to be upset about
something, it will leap into the air to the full extent of the

jesses and leash, and then fall until it is hanging upside down from your hand, projecting a sulky refusal to play any constructive part in the proceedings. Wellington was in no pain or discomfort; he could perfectly well get back up by himself with a single beat of his wings—as he would rapidly demonstrate if he ever fell off a perch by accident. But no; there he hung, twisting gently by his ankles, wings half open, obstinately refusing to see reason. When you are not used to this behaviour it naturally seems alarming—you are concerned that the bird may injure itself. When you have picked it up and sat it back on your fist twenty times in an hour, only to be rewarded with another *kamikaze* dive, you tend to get irritable. If you give up and put the dratted creature back on its perch, you lose ten points.

With effortless ease, Wellington invariably beat me at this game. He was not going to take food from my hand; he was not going to let me stroke him; he was not even going to sit on my fist for more than a moment at a time. As a creature designed by evolution for solitary hunting by means of untiring watchfulness, he had infinite resources of patience. I—being designed to catch my dinner by dashing around the savannah with my mates, in a state of noisy excitement—did not.

◉◉

DURING THAT WINTER of 1977–78 I had to go away for a week-long business trip, and Dick kindly agreed to put Wellington up in a vacant aviary while I was away. When I returned and rang to arrange a trip to collect him again,

Dick's voice was apologetic. He told me that while I was away Wellington had found a chink in a corner of the mesh and had escaped into the night.

My response to this news was distinctly mixed. Firstly, I was sorry to have given Dick reason to worry. Secondly, I was sorry that Wellington had put himself in serious danger. It was not only that he had no experience of hunting for his food; more important, he was wearing jesses and a swivel. This meant that he ran a serious risk of getting himself tangled up in the branches of a tree, with fatal consequences. I could only hope that he would survive long enough to gnaw through one jess, which would at least leave him trailing a single length rather than a more dangerous loop. But thirdly, I was frankly relieved at being freed from a failing project without actually having had to make a decision.

It would have been ridiculous to claim that I was fond of Wellington. He had been a prisoner and I his jailer; we had achieved no other relationship, nor was there any hint that we ever would. There was nothing for it but to write the whole thing down to experience, and get on with my life.

In the event, it didn't turn out like that. As the weeks passed, the lack of focus to my evenings at home became faintly disagreeable. The empty cage on the balcony—in front of my eyes outside the office window whenever I sat down at the typewriter—was a frequent reproach. My original wish to own and tame a bird of prey had been dampened but not quenched, and I couldn't help brooding about the gap between what I had wanted to achieve and what had actually happened. During the family Christmas holiday at Water Farm the benign and fluffy presence of Wol, calmly presiding over the festivities from his shaded perch high up in a corner of the kitchen, was a constant reminder. In the New Year, I admitted to myself that I still wanted an owl. But what sort of owl?

◎◎

AFTER WELLINGTON, I didn't want another *Athene noctua*, but there would be little problem in acquiring a Barn Owl— *Tyto alba* (of which there are many more in captivity these days than in the wild, some of them birds rescued after being injured). The Barn Owl is the glamour puss of TV natural history programmes, which seem to queue up to film them inside nesting-boxes. The Latin name means White Owl; it is sometimes also known as the Screech Owl, from the blood-chilling shriek that occasionally scares the pants off night-time dog-walkers in the countryside. (In the United States, Screech Owl is the name of a different species.) Obviously, it takes its everyday name from its habit of nesting in farm buildings and setting up a mutually beneficial hunting territory in farmers' yards and fields.

The Barn Owl's stiffly heart-shaped face gives it a haughty look of self-conscious dignity, and its magnificent golden-buff and white plumage dusted with dark speckles makes it highly visible (sometimes startlingly so, given its silent, ghostly approach) as it quarters its territory during hunting patrols both by night and in daylight. These photogenic looks, coupled with its precarious population numbers, make it the poster-owl for conservationists, and its flexible daily timetable and willingness to live close to humans are convenient for wildlife photographers and film-makers. But beautiful and graceful though the Barn Owl undoubtedly is, I must confess that I have never really warmed to it emotionally, and, while it thrives in aviaries, the books told me that it seldom makes a satisfactory house-pet.

The Tawny Owl—*Strix aluco*—studiously avoids people with film cameras and PhDs, and lives a much more private life than *Tyto alba*. It does not patrol on the wing, is much less tolerant of humans nearby, and spends most of its time sitting motionless and camouflaged in woodland trees. Logically, it should therefore seem a more remote and less sympathetic creature, but human sentimentality doesn't work that way. When confronted with a tawny, it would surely take a heart of stone not to feel the urge to cuddle it. It is well known that humans instinctively respond most warmly to animals (especially young ones) with soft fur or feathers, and faces in the place that we expect to see them. This "*aaah* . . . factor" is, of course, just sentimental anthropomorphism, but its force is so undeniable that it seems pointless to fight it.

We identify with owls because they have an upright stance and a recognizable face. Tawnies also have a rounder head than Barn Owls, and a rounder facial disc that is more softly delineated and less "aristocratic." Their dark eyes are relatively larger and, unlike those of Barn Owls, are not widely separated by a smooth, vertical ridge of protruding feathers. Instead, a widow's peak of short, contrast-coloured feathers growing down from the "forehead" seems to cut the upper edge of the facial disc into two "eyebrows." We automatically perceive the short, hooked bill as a "nose," and the corners of the mouth are hidden by a broad triangle of "whiskers."

Watch a tawny for any length of time, and the subtle movements of the tiny, dense feathers covering its face give it what we perceive as a more expressive appearance than the rigid, heart-shaped mask of the Barn Owl. These changes do not, of course, have any real correlation to human facial expressions of emotion, but they look as if they should do—in the same way that a panting dog can look as if it is grinning. In repose, hunkered down and fluffed up into the shape of a cottage loaf, tawnies have a contentedly sleepy, comfortable look, and their brown and off-white plumage gives an illusion of being softer than that of the sleeker-looking Barn Owl. And finally, the tawny has the reputation of making a happy pet.

After the briefest mental flirtation with thoughts of a majestic (and, more to the point, usually silent) *Tyto alba*, my choice was easy; I wanted a tawny myself.

∞

IN BRITAIN ALL wild raptors and their eggs and nestlings are strictly protected by law. I discussed my plans with Dick, and phone calls were made to knowledgeable people with—in this case—the correct legal paperwork. In due course word came back, with a price for reserving the first egg laid in the coming spring by a hen tawny in the aviaries of a licensed breeder of birds of prey. I took the plunge, and placed my order.

This choice of egg was apparently important. In captivity—where conditions replicate years of abundant prey in the wild—most tawnies lay their eggs in clutches of up to five. Just as in the wild, these are laid in a staggered sequence at intervals of at least two and sometimes several days. The owlet that emerges from the first egg to hatch, about four weeks later, has an immediate advantage over its fellow nestlings. For at least a couple of days and often longer, it enjoys the parents' exclusive attention and feeding. By the time its siblings hatch, it is bigger, stronger, and more loudly insistent in calling for food, and so has a better chance of thriving.

In the wild this parental favoritism is particularly marked in years when prey is scarce; it gives the greatest chance that at least one robust hatchling will survive in good shape to face the huge challenges of fending for itself during its first autumn and winter. Tawny owlets may look adorable (unlike the rather vulturine babies of the Barn Owl), but any raptor's nest is as mercilessly competitive an arena as the wild woodlands, and it is quite common for one or more of the later hatchlings to perish there. Nature being Nature—and the appetite of hatchlings being insatiable—these unlucky early losers in life's race are not allowed to go to waste.

There also seems to be some folklore about the first egg usually being female, since female tawnies grow bigger than males. This question of gender is problematic, since there is no way of "sexing" a young owl without intrusive and very expert examination, or even subjecting it to X-rays. In common with other birds, owls have no external genitalia, and while a mature hen bird may weigh 25 per cent more than a cock, young tawnies show no obvious gender differences in size or coloration. (Although I decided to assume that my tawny was female, it was in fact only when she was a couple of years old that I became convinced, by a particular behaviour, that she was indeed a "she." But it would be ridiculous to describe her as "it" up to that point in this story, so "she" she is, throughout.)

◎◎

EARLY IN APRIL 1978 the report came through that "my" egg had been laid, removed from the nest, and placed in an incubator. About four weeks later I was told that the owlet had hatched successfully, and was being hand-reared by the breeder's small son. During these first weeks a hatchling needs repeated hand-feeding, several times a day, with something fairly gloopy on the end of a matchstick. I was further informed that when the little boy had asked his dad what he should call the owlet, and had been told to name it after something that he really liked, he thought for a moment, and then announced that its name was "Marmite Sandwich."

At last the day came, on a gloriously sunny Saturday late in May, when I drove down to Water Farm to begin the second chapter of my involvement with owls. It was a thoughtful drive. A small, doubting voice from a joyless corner of my mind kept asking me if I really knew what I was letting myself in for, again—the mess, the inconvenience, the complication to my social life, and all probably leading only to another disappointment. Why not be pragmatic, and simply admit that I wasn't an animal person? Usually the swoop down Bluebell Hill with half of Kent laid out below me was enough to raise my spirits on the gloomiest day, but on that sunlit Saturday I never noticed it at all.

One of the relaxing things about arriving at Water Farm was the lack of reaction it usually caused. There was no reception committee; you tended to encounter various members of the family one by one, scattered about the place busy with their own concerns, whether having a mug of coffee

under the willow tree beside the duck pond, or doing something rural to a sheep's feet in the paddock. In time, a pause
in the sound of grinding and drilling from behind the old
tractor shed would signal that Dick was taking a break from
performing open-heart surgery on something rusty that
had been written off by the U.S. Army in 1945. I always
found this interval restful, because it gave me time to get
the motorway hum out of my brain and readjust to human
conversation. That Saturday my arrival caused the usual lack
of excitement, and there was nothing at all to warn me that
I was walking towards a momentous encounter.

Me: "Where is it?"

Dick: "In the kitchen."

The door straight into the big farmhouse kitchen stood
open, as usual, and I stepped into the cool shadow. Wol,
dozing on his perch high up on top of the dresser, ignored
me. I looked around, expecting to see the usual cardboard
grocery box somewhere, and to hear the usual fright-
ened scuttling of claws inside it as I approached. No scuttling; no box.

Perched on the back of a sunlit chair by the open window was something about nine inches tall and shaped rather
like a plump toy penguin with a nose-job. It appeared to be
wearing a one-piece knitted jumpsuit of pale grey fluff with
brown stitching, complete with an attached balaclava helmet. From the face-hole of the fuzzy balaclava, two big,
shiny black eyes gazed up at me trustfully. "*Kweep,*" it said
quietly. Enchanted, I leaned closer. It blinked its furry grey
eyelids, then jumped very deliberately up onto my right
shoulder. It felt like a big, warm dandelion head against my

"Something about nine inches tall and shaped rather like a plump toy penguin with a nose-job. It appeared to be wearing a one-piece knitted jumpsuit of pale grey fluff . . ."

cheek, and it smelt like a milky new kitten. "*Kweep*," it re-peated, very softly.

◎◎

ON SUNDAY EVENING we drove back to London together. There had been no need to ask Dick's help in fitting jesses; I had sworn to myself that this bird would never be tethered, and would come to me of her own free will or not at all. The owlet started the journey in a cardboard box, but soon es-caped and climbed up to my shoulder. She adapted herself to this revolutionary new experience with complete calm. By the time we were half-way home she had learned to lean into the corners for balance, occasionally taking a delicate beak-grip on the rim of my ear to steady herself.

Love at first sight—when it hits you late, it hits hard. It hit me at thirty-four, and I was a slave to it for the next fifteen years.

2

◎◎

Owls—the Science Bit,
and the Folklore

I N T E R M S O F traceable evolution, my Tawny Owl sprang from a hugely more ancient lineage than you and I, and if she had had room in her close-packed little skull for speculative thought she would have had every right to look down on humanity as pathetically slow learners.

We know much less about the evolution of birds than about that of most other classes of animals, because their flimsy bones rarely survive as fossils, and even those that do are extremely difficult to identify. It is accepted that birds originally evolved from small reptiles during the age of the dinosaurs, though the process by which this happened remains a matter of academic debate. The oldest piece of evidence for this specialization has long been recognized as *Archaeopteryx*, of which a group of several sharply delineated fossils were found in Germany. These have been dated to the Upper Jurassic period, about 160 million years ago (when the age of the dinosaurs still had about 95 million years to run). The crow-sized *Archaeopteryx* has a reptilian skull with toothed jaws but a skeleton that shows both reptilian and avian characteristics; most strikingly, it undeniably has feathered wings, but also a long reptilian tail fringed with feathers. (Recently, a fossil found in China and christened *Xiaotingia*, showing long feathers on all four limbs and the tail, has been claimed as an even older creature.) But the subsequent series of adaptations that linked these creatures with today's birds—if, indeed, such direct links exist—represent a vast jigsaw puzzle of which we have identified only a very small number of the pieces. Evolution is a process punctuated by unknown numbers of dead ends, and many known fossils are those of species that later became extinct without leaving any identifiable descendants.

However, we do know that when our own most distant probable ancestor, the proto-hominid *Australopithecus afarensis* (familiarly known as "Lucy"), emerged onto the savannah in the Horn of Africa at some time around 3.5 million years ago, the dawn-mother of the owls, *Protostrix*, had already been part of the fossil record in what is now North America for at least 50 million years. *Protostrix* had appeared during the Eocene era, long before the tectonic plates that underlie our continents had separated and slid into their present positions. The last of the dinosaurs had died out some 15 million years before *Protostrix* laid down her imprint in the shale-mud, at a time when both the birds that had evolved from some of the smaller reptiles, and the small mammals that these birds preyed upon, were diversifying and spreading through the forests and grasslands that covered much of the planet.

Unimaginable periods of time then passed while evolution carried out its infinitely slow series of branching experiments, filtering the genetic legacies of the flying hunters. By the Pleistocene period, "only" some three million years ago, this process had produced something that was definitely a Tawny Owl. *Strix aluco* is one of about thirty of the most ancient species of the family *Strigidae* that are still recognizable today, among many more recent arrivals. That the Tawny Owl continued to survive the ceaseless process of competitive species-sorting argues that by three million years ago it had already adapted well to its environment, becoming the master of a particular niche of opportunity in the food chain. This, remember, was at a time when our own hairy, dwarfish ancestors were still coming to terms with walking on their back legs, and were still at least half a million years from discovering the potential benefits of deliberately banging rocks together.

The exact sequence of relationships in mankind's family tree remains a matter of debate, since a number of the trial-and-error sketches for a future human being that have been revealed by fossil finds seem to overlap in period rather than representing successive stages in the process. About a million years after Lucy, some 2.3 million years ago, a still ape-like creature called *Homo habilis* had a brain about 50 per cent larger than hers, and was definitely using stone tools. By 1.8 million years ago *Homo erectus*, the first undoubted member of our own lineage, was on the scene in eastern Africa—taller, straighter, less furry, and with a brain that continued to grow in size, slowly but steadily, over hundreds of thousands of years; the rock-banging was now getting more ambitious and sophisticated. At some unknown time a number of families of this upwardly mobile species migrated from Africa into the Middle East and began their slow colonization of the rest of the planet.

From about 800,000 years ago the enlargement of *Homo erectus*'s brain speeded up remarkably, presumably in response to the challenge of dramatic and repeated changes in climate and environment. By just 120,000 years ago (in evolutionary time, the blink of an eye), we—*Homo sapiens*— were jostling our way ahead of the competition to challenge the brawnier and equally large-brained *Homo neanderthalensis*, but it was a mere 28,000 years ago when the last

Neanderthals disappeared from ice-bound Europe. (What gave us the edge may have been as simple as the bone needle: apparently we had it, so we could sew hides into warm clothing, but Neanderthals didn't.) As a parting souvenir of interbreeding they bequeathed us about 4 per cent of our DNA, but they left us as the final champions in the long marathon of primate evolution. We now had a brain more than three times the size of Lucy's, a truly upright stance, a skeleton evolved for long-distance running, strong opposable thumbs, and a gut that could cope with a wide variety of foodstuffs. Together these gave us such unparalleled adaptability that—though comparatively weak, and pathetically slow to leave our mothers and breed—we were able to survive or sidestep many of the basic daily challenges that mercilessly cull each generation of our fellow inhabitants of the Earth.

Meanwhile, throughout this whole glacially slow process, Tawny Owls had been contentedly staking out territories, catching their supper, and making baby Tawny Owls. Unlike us, they had not had to adapt in any major way; working from a fifty-million-year-old basic template, Nature finally got them right three million years ago, producing a magnificently structured and equipped killer living in complete harmony with its environment. So long as the Earth still has breathable air, killable prey, and perching-places well above ground, they could theoretically carry on being Tawny Owls for eons into the future. (In the narrow terms of comparative evolution, they might therefore be characterized as a "dead end"; but their inability to compose orchestral symphonies or manufacture thermo-nuclear weapons does not seem to handicap them much.)

◉◉

TO A CURIOUS child's reasonable question "What are owls *for*?" the easiest answer is that they are something like cats that can fly, which enables them to share the cat's work at night. (There seems to be a weird sort of rightness in Edward Lear's pairing of the two creatures in a loving relationship in his nonsense-poem "The Owl and the Pussy-cat.") In the infinitely interconnected web of life on Earth, owls fill the job opportunity that occurs every night when the day-time hunting birds go to sleep. These night-shift workers are an essential part of the mechanisms that keep the population numbers of the Earth's creatures in some sort of sustainable balance. We might even tell the child that without owls, farmers would be up to their necks in rats and mice.

More seriously, when the child gets older and is still curious, we might explain what it is that sets an owl apart from other birds. The defining characteristics of the avian order of *Strigiformes* are large, forward-facing eyes—more than twice as big as those of other birds of comparable size—and very well-developed ears, which together enable them to hunt live prey in what we would (mistakenly) call absolute darkness.

The different families, genera, and species of owls all share a strong, compact body generously feathered with dense, soft plumage that hides a long, flexible neck, powerful legs, and big hooked talons. They also share a habit of immobility; they spend most of their time sitting quietly and watching the world go by, so they have no need for any vivid coloration—camouflage is more important to them than advertising their identity from a distance, or putting on scary deterrent displays (though they can do that too, when they need to). Because of their unobtrusive colouring and their nocturnal way of life, many species—especially those that favour thick woodland rather than open country—depend on sound for communication. They are much more vocal than most daytime raptors, and they have a wide vocabulary of specific calls.

◉◉

ABOUT 135 SPECIES of owls, grouped in 24 genera, have been classified the world over. ("About," because new species are still occasionally discovered in remote jungle regions, and also because taxonomists disagree over whether or not some sub-species should properly be classified as truly separate.) The great majority belong to the family *Strigidae*, but about ten species are classified as *Tytonidae*—the Barn and Grass Owls, which share some distinct anatomical differences from the rest. Ornithologists have made quite deep studies of a limited number of the world's owl species, but their habitats and nocturnal behaviour make them difficult subjects for field observation, so, beyond their mere existence and some basic characteristics, many species are still very little known. One thing is certain: owls have adapted successfully to all but one of the almost limitless range of environments with which they have been confronted by the millenial cycles of climate change and species proliferation.

In size, they range from the several species of Eagle Owls—which come up to your hip when you stand beside one that's sitting on a low ground perch—down to Elf and

Pygmy Owls that are the size of small garden songbirds. The very biggest is said to be Blakiston's Fish Owl, some of which reach 30 inches tall and weigh 10 pounds, with a wingspan of 6.75 feet (2.05 metres). The smallest, the Elf Owl of the south-west United States, is about 5 inches (13 centimetres) tall, and weighs around 1.5 ounces (42.5 grams)—one-hundredth the weight of an Eagle Owl.

Owls are found on every continental landmass except Antarctica, and on many remote islands. Most of them live in or on the edges of forest, in regions as diverse as the frozen sub-arctic taiga and tropical rainforest or swampland. However, some species also thrive in treeless terrain: owls can be found living on open prairies, in arid deserts, and on the Arctic tundra. Most species nest in trees or on rocky ledges, but others on the ground, or even in underground burrows. Many species remain on their home turf all year round, but some will disperse more widely when local prey becomes scarce, and others regularly make seasonal migrations over land and sea. As a rule of thumb, the more varied an owl's appetite, the more likely it is to stay in one place, and to form a long-term bond with a mate; the more specialized its diet, the more likely it is to wander in pursuit of that prey—like nomadic human hunters following the herds—and to form only relatively brief relationships in the breeding season.

Only perhaps 40 per cent of owl species are strictly "nocturnal" (that is, they hunt only between nightfall and dawn), while many others are active both in the light evenings and by night. A number of species—and not only those that have adapted to the "white nights" of high summer in the Far North—hunt in full daylight. Many owls hunt by the "perch-and-pounce" method: they watch patiently from lookout points until they spot a potential meal below, then dive down to make their kill. However, some species (especially among those that are active in the daytime and dusk) hunt on the wing, like hawks and falcons, and others may run around chasing prey on the ground.

Their diets range all the way from insects (including the whole menu of what we might simply call "creepy-crawlies"), through invertebrates such as worms and slugs, to snakes, crustaceans, frogs and other amphibians, up to rodents small and large, rabbits and hares, cats, dogs, foxes, and occasionally even young deer. Many routinely prey on other birds, ranging in size from sparrows up to herons; in coastal areas the largest owls have been known to tackle even such dangerous fellow predators as the Great Skua, and in northern forests large owls regularly hunt smaller owls. Some African and Asian owls specialize in fishing, and in the rivers of Siberia and northern China it is common for Blakiston's Fish Owl to snatch prey as large as salmon and pike.

The relative numbers of the different types of prey taken by any particular species of owl vary with the regular cycles of availability of each edible creature in their hunting territories. These cycles, which recur over a few consecutive years, also govern the breeding numbers of owls, and thus their own population density.

◎◎

FIVE SPECIES OF owls are normally resident in mainland Britain: in descending order of population numbers, the Tawny, the Little, the Barn, and the Short-Eared and Long-

Eared Owls. Additionally, the great white Snowy Owl has been known to breed during summers spent in the Shetland Isles, but is otherwise only a winter visitor to Scotland from its sub-arctic home ranges in Scandinavia. There are marked differences—in habitat, in timetable of daily activity, and to a lesser extent in preferred prey—among the ecological niches occupied by the different species, and this allows them to coexist without too much direct competition.

The Little and Barn Owls have already been described briefly. The scarce and closely related Short-Eared and Long-Eared Owls are not found in southern or eastern Britain, but mostly in the north and west. (Their "ears" are in fact simply tufts of feathers on their heads that are used for recognition and signalling, and have nothing whatever to do with their actual ears.) The populations of both species are notoriously uncertain, but there may be about 3,500 breeding pairs of Short-Eared Owls in Britain (so slightly less than the Barn Owl population), and fewer than 1,000 pairs of Long-Eared. Both are migratory species, and in autumn and winter Scandinavian visitors increase the numbers of Short-Eared Owls. Birds that migrate seasonally are less solitary in their habits than those that stay close to home. Both these species are less territorial and—except in the breeding season—more gregarious than the solitary Tawny, Little, and Barn Owls. Small squadrons of Long-Eared Owls have been seen migrating from Scotland southwards and westwards in autumn, and groups of them will roost together in wintertime. So long as prey is plentiful, different pairs of Short-Eared Owls will tolerate each other when living quite short distances apart.

These two species are roughly similar in size and

colouring, but very different in their habitats and ways of life. The Short-Eared are birds of open heath, grassland, and marsh, which nest and spend much of their time on or close to the ground. With relatively long wings, and lemon-yellow eyes, they hunt on the wing by day and at dusk. Long-Eared Owls are strictly nocturnal woodland birds, found particularly in conifer forests; they have relatively short, broad wings and orange eyes, and hunt on the wing along the edges of their home forest. Always sparse in numbers, but formerly found distributed over wide areas, Long-Eared Owls suffered a marked decline in Britain in the twentieth century. Simultaneously, an owl of similar size and colouring but much more adaptable habits began to extend its range into their territories: *Strix aluco*, the Tawny or Brown Owl.

◉◉

SO MUCH FOR the paleontology and zoology; but what about the "sociology"—historically speaking, how do we feel about owls?

Ancient cave paintings in France and other surviving images from around the world confirm that mankind's conscious relationship with owls stretches back over tens of thousands of years, and they figure more often than any other bird in human myth and folklore. Puzzlingly, our feelings towards them have always been remarkably ambivalent: humans have felt respect for owls' actual or imagined qualities, while simultaneously regarding them with superstitious dread.

The practical aspects of mankind's interaction with owls

have mostly been positive. Throughout most of our history we have never seen owls as competitors for food resources—indeed, we have recognized that they are actively helpful to us. Since the birth of agriculture some ten thousand years ago farming has overwhelmingly meant the cultivation of grain crops, and rodents have always been the scourge of the grain-farmer—they plunder his harvest, foul his stored grain, and spread disease. Being more versatile than cats, owls are Nature's greatest killers of rodents, so having owls around the place was an unambiguous benefit for farmers. In parts of northern Europe you can still see both field perches deliberately provided for them and traditional farmhouses built with pierced "owl-boards" in the gables to encourage nesting Barn Owls, and in several of those countries folk-wisdom has been reinforced by legal protection. (Historically, captive owls were also used by bird-catchers to lure "mobbing" birds into nets, or onto nearby twigs smeared with gluey bird-lime.)

Nevertheless, superstition has influenced human attitudes towards owls much more powerfully than any commonsense recognition of their usefulness. On the positive side, in Western and some other cultures owls have been associated with wisdom (though interestingly, they have the opposite reputation in Indian folklore). Europeans have always been impressed by the owl's serene stillness during daytime: our forebears assumed that anything that spends so much time sitting quietly and keeping itself to itself must be deep in thought—thus, the "wise old owl," who watches everything but says nothing. The ancient Greeks associated the Little Owl, which is common in Mediterranean countries, with their warrior goddess of wisdom, Pallas Athene.

Depictions and literary allusions to the goddess often included owl imagery, and since she was the tutelary deity
of the city-state of Athens, her Little Owl even appeared on
some Athenian coins. Among more distant cultures from
our own there were a few others that positively revered the
owl, such as the Mongolians and Tatars. Some Native
Americans believed that the souls of their medicine-men
passed into owls, and in southern Australia the aboriginal
peoples regarded them as, specifically, the guardian spirits
of women.

Since literacy and learning in medieval Europe were
almost exclusively the province of the Church—and since
owls often nested in church towers—owls became popu-
larly associated with the clergy. A twelfth-century English
allegory, *The Owl and the Nightingale*, mentions both Barn
and Tawny Owls—respectively, "the owl that scritchest" and
"the owl that yollest." (When the author of the illustrated
Ashmole Bestiary of the early thirteenth century comments
on what is clearly a Barn Owl that it is "heavy with feathers,
signifying superfluity of flesh and lightness of mind," he is
simply showing his ignorance; an owl's flesh is anything but
"superfluous.") In the Arthurian myth cycle the wizard
Merlin was supposedly accompanied by an owl sitting on
his shoulder. The owl's sedentary stillness, cloaked posture,
and big eyes surrounded by encircling feathers would later
remind people of a bespectacled scholar or schoolmaster—
respected for his learning, if mocked for his stuffy dignity
and unworldliness.

In all superstitious societies, owls' body-parts have
figured in the rituals and recipes devised to summon up
"sympathetic magic," and such beliefs have ranged from the

straightforward to the fanciful. It is easy to understand why Apache warriors in the American South-West decorated their war-caps with owl feathers, invoking the skills of silent, stealthy hunting. A rather more plonkingly literal approach persuaded several other peoples to eat owls' eyes in the hope of improving their night vision. In Yorkshire, owl soup was supposed to cure whooping-cough, and even in early modern times there was an English country superstition that feeding a child the egg of this "sober" bird would protect it from growing up to become a drunkard. (Again, Indian folklore went its own way: in India, owl flesh was imagined to be an aphrodisiac.)

◎◎

THE PLACE OCCUPIED by owls in our folklore has reflected a confusion between respect and fear, but heavily skewed towards the latter. Unfortunately, the negative images of owls have always far outweighed the positive, and this is obviously due to their association with the night. Night-time for humans meant blindness and helplessness in the face of real or imagined terrors. Night was when ghosts and evil spirits walked abroad, and a creature that was specifically a master of the night must therefore be a consort of the dark powers.

Only a few centuries ago in Europe owls were one of the animals habitually supposed to be the devilish "familiars" of witches; for instance, the *Book of the Days* reporting the trial of the three "Belvoir witches" in Leicestershire in 1618 has a woodcut including one of them, Joane Willimot, with an eared owl on her shoulder. This reputation as a witch's accomplice was not simply a case of the owl providing an unwitting conduit for evil. It was argued that any bird that hid from the sunlight and was habitually abused—that is, "mobbed"—by daytime birds must bear the curse of some ancient crime. The Old Testament lists the owl among the creatures to be abominated, and as the inhabitant of ruins it was also a dreary reminder of the vanity and decay of human hopes: "It shall never be inhabited . . . But wild beasts of the desert shall lie there; and their houses shall be full of doleful creatures; and owls shall dwell there . . . And thorns shall come up in her palaces, nettles and brambles in the fortresses thereof; and it shall be an habitation of dragons, and a court for owls" (Isaiah 13:20–21 and 34:13).

By extension from its occult associations and its dark and eerie haunts, the owl became—above all other animals— the creature of ill-omen, and the herald of misfortune and death. Oddly enough, while the Romans were great respecters of Greek culture, and identified Pallas Athene with their equivalent goddess, Minerva, their feelings about her symbolic bird were overwhelmingly negative (although, by contradiction, images of owls were supposed to counter the "evil eye"). Pliny the Elder's *Natural Histories* are a hilarious repository of utter nonsense about animals and the remedies to be derived from their body-parts; his ignorance is exemplified by his belief that owls had defective vision, and his credulity about their malign supernatural power seems to have been widely shared. In his society, whose priests interpreted the behaviour of birds when studying the auspices before public decisions were made, "owl" was already a slang term for a witch. An archaic translation of Pliny declares, "The scritch-owl betokeneth always some heavy news, and is most execrable and accursed in the presaging of public affairs." To Pliny, the owl was "the very monster of the night," and the sound of one "foretells some fearful misfortune" (though history does not record whether he heard one before his fatal boat trip to examine the eruption of Vesuvius in A.D. 79).

Shakespeare echoes this Roman superstition in his *Julius Caesar*, in which the dictator's assassination is foreshadowed by an unnatural daytime appearance: ". . . the bird of night did sit / Even at noonday, upon the marketplace, / Hooting and shrieking." He has Lady Macbeth call the owl ". . . the fatal bellman, / Which gives the stern'st goodnight." Elsewhere he writes that its ". . . scritching loud / Puts the wretch that lies in woe / In remembrance of a shroud" (*A Midsummer Night's Dream*), and even "The owl shrieked at thy birth, an evil sign" (*Henry VI*). A contemporary Austrian source describes the owl as "*ein Totenvogel, Sinnbild der Sünde*" ("a death-bird, symbol of sin"). Again in the sixteenth century, Edmund Spenser calls it "death's dreadful messenger," and Robert Jones writes: "Come, doleful owl, the messenger of woe / Melancholy bird, companion of despaire." In fact, English poets from Geoffrey Chaucer in the fourteenth century (who called it "the prophete of wo and myschaunce") right up to Edward Thomas and Laurie Lee in the twentieth have been shamefully negative about this innocent and useful bird. In 1808 Oliver Goldsmith went so ridiculously far as to call owls "nocturnal robbers," and complained that to hunt by night was simply unsporting!

The belief that hearing an owl calling on the roof or even nearby was a sign of an impending death in the household ("and sings a dirge for dying souls"—Thomas Vautor, c. 1600) seems to have been almost universal among country folk for many centuries, and presumably originated in the simple fact that most natural deaths occur at night. The Chinese went further and believed that owls actually snatched away the souls of the dying, while there was an Arab tradition that owls embodied the spirits of unavenged victims of murder, crying out for blood. (More prosaically, it seems that in Wales the owl's cry foretold another event that usually happened at night: it was a warning that a maiden was about to lose her virginity.)

Much of this might seem to answer the belligerent demand that I recall making to somebody whose apparent disapproval of my own domestic arrangements was becoming irritating: "So, what have you got against owls anyway, chum?" But I would naturally object that Mumble's forebears suffered an undeservedly bad press, and even today her fellow owls may sometimes continue to be slandered, if for rather different reasons.

◉◉

IN EARLIER TIMES, when mankind came into direct contact with predatory carnivores that took their livestock and might threaten their children, there were good reasons for feeling hostility towards them. If you lived in a shack in a lonely forest clearing, or depended upon your small flock for survival, you could be forgiven for not giving wolves or eagles the benefit of the doubt.

In modern times most of us have lost all instinctive understanding of the checks and balances of the natural world, to the extent that many people have become babyishly squeamish about the simplest facts of animal life. If there are adults so ignorant that they honestly do not connect the plastic-wrapped slab of supermarket meat with the idea of a living cow, then it is not surprising that they prefer not to think about the indispensable role of predatory animals in the universal cycles of life. For instance, even those who enjoy an opportunity to see birds of prey on the wing may exclaim with prissy indignation about their "cruelty" when they see one stoop for the kill, and imagine the muffled squeak of a small life ending.

In fact, of course, among the birds of prey, owls may get something of a free pass in this respect; they look cuddly, they hide their killing talons under fluffy feathers, and they are only out about their natural business when most people are asleep. (The owls in the "Harry Potter" films are seen benignly carrying messages to the dining tables of Hogwarts—audiences are unlikely to give a thought to what they get up to at their own mealtimes.) Moreover, the almost complete urbanization of western Europe has made actual encounters with wild owls fairly rare. Since most British people now live in towns and cities, they will never be startled by the silent, ghostly-white swoop and bloodcurdling shriek of a patrolling Barn Owl.

However, the fading of religious belief certainly has not made us less susceptible to ghost stories. After dark, our rational minds cannot always protect us from folk memories about these haunters of ruins and graveyards who come as winged messengers of death and calamity. For some people, even the much less alarming hoot of a Tawny Owl as they walk past a dark wood makes their neck-hairs prickle if they are not accustomed to being out of doors in the countryside at night. That mournful-sounding call seems to reinforce feelings of loneliness, and of near-helplessness in a surrounding darkness that might hide the approach of unknown dangers.

◎◎

FOR OBVIOUS REASONS, my own reaction to that sound has been very different ever since a Saturday afternoon in May 1978.

One summer, after Mumble had passed out of my life, a friend and I decided in a spirit of curiosity to try sleeping out in an ancient Hampshire yew wood that was notorious for being haunted. Supposedly, it is the site of the thousand-year-old mass grave of the victims of a battle between Vikings and West Saxons. I was told by somebody who had grown up near Chichester that local children used to dare each other to venture into it, and a botanical artist had told another friend of mine that she suffered a panic reaction when she walked through it after being caught out unexpectedly late. Even the warden of the nature park in which this unusual concentration of yew trees grows—a thoroughly practical outdoorsman—has admitted in print that he avoids it as soon as the sun begins to dip below the horizon, claiming to be conscious of a sinister atmosphere under its dark eaves.

Despite this evil reputation, Will and I didn't have much hope of experiencing anything interesting. Nevertheless, after an evening at the nearest pub we trekked in with sleeping bags and water bottles and settled down among the ancient, riven trunks and the caves of ground-sweeping branches. Will, a longtime Territorial Army rifleman, efficiently scraped a small pit for his hipbone, slithered inside his "green slug," and was gently snoring within minutes. Less accustomed to roughing it, I lay awake for a long time, listening to the breeze and the occasional faint patter of drizzle. My emotional antennae were consciously tuned to pick up any chilling sensations, but I was completely disappointed. As my eyes at last grew heavy there suddenly came, from about two trees away, the tremulous call of a Tawny Owl. *Aaah, sweetheart* . . . I thought; and I rolled over to sleep, comforted like a child and feeling foolish about the whole ghost-hunting exercise. In my world, nothing wicked could be abroad if there was a tawny nearby.

3

◎◎

The Stowaway on the Seventh Floor

SINCE IT WAS the last week of May 1978 when I collected "Marmite Sandwich" from Water Farm as a fledgeling roughly thirty days out of the egg, she must have hatched in late April. I decided that, like Her Majesty, my owl should have an "official" birthday. Because I have a connection with the British old comrades' association of the French Foreign Legion, I took the entirely arbitrary decision to celebrate it on the auspicious 30 April—the Legion's "Cameron Day," when I would be at a party anyway.

My owl's legal identity was "39 RAH 78 U," stamped in black on a yellow plastic bangle worn discreetly around her feathery left ankle. The name "Mumble" just came into my head after a few days of listening to her quiet conversations with herself, me, and the world at large. (I must emphasize that this all happened about thirty years before a major American animation studio gave the same name to a fictitious penguin with showbiz ambitions. I have always been vaguely suspicious about that, but I don't recall that anyone from Hollywood ever met my owl.)

◉◉

BEFORE I BROUGHT Mumble home I had cleaned out Wellington's old balcony cage, and after some thought I had also constructed a second, indoor cage on a worktop against the window in my kitchen. I was not sure how our routine was going to work out, but after my experiences with Wellington it seemed sensible to prepare for a certain amount of flexibility in the domestic arrangements. I modified a kit for making garden compost bins—wide-mesh panels of plastic-covered wire grill, fixed together with snap-on rubber clips. This produced an airy metre-square cube; I installed a door at the front fitted with a "doorstep perch," a couple of sawn-branch perches in the rear corners, and the usual thick carpet of newspapers on the floor.

I was determined never to tether Mumble, and I wanted to give her the freedom to spend plenty of time loose in the flat—or at least in the circuit provided by the hallway, bathroom, kitchen, and living-room (common sense suggested barring her from my bedroom and office, though I would occasionally have moments of weakness—she could twist me round her little claw). Since it is impossible to housetrain a bird, that meant that I would have to accept natural accidents with a degree of equanimity: in short, that there was bound to be a fair amount of owl-crap around the place, and I had better get used to the idea.

Tawnies are cleanly birds compared with Barn Owls, and almost never foul their nesting-places, but this means that they let fly when they are roosting. Although I knew that I would never be able to persuade her to defecate in one predictable place, I did think it worthwhile to provide her with as invitingly comfortable a facility as possible. I built a large tray-perch by fixing a sturdy, L-shaped "gallows" branch upright on a big wooden bread tray that I spotted in a Dumpster behind a supermarket, lined the whole thing generously with newspaper, and set it up on a small table in the living-room with a good view of the windows. (For some long-forgotten commercial reason the bread tray had the word "Perfection" stencilled boldly along one side. When Mumble chose to sit in fat contentment immediately above this caption, the impression of upholstered mandarin conceit was striking.)

I knew that when she was free to choose she would prefer to sit high. The most obvious perches would therefore be on the tops of half-open doors, and it would be sensible to cover the floors around these with newspaper. I also taped a good many thin plastic sheets around the place, to protect walls from "splash-back" and to cover furniture that was seldom used when I was alone in the flat. This routine was going to be demanding, but that was an inevitable price for sharing my quarters with a wild creature.

It was obvious that I would once again have to recruit my next-door neighbour, Lynne, as an essential ally, since Mumble was going to be spending a lot of time on the balcony only a few feet from her bedroom window. When I hesitantly explained to her that I was planning to acquire a potentially songful flatmate she took it amazingly well— and would continue to do so, throughout the whole time she had to put up with Mumble's occasional serenades. After a few months she moved out and sub-let her flat, but the reason was marriage rather than owl-fatigue, and her tolerance would be inherited by a couple of her subsequent tenants. (One of these was my old friend Gerry, an illustrator with whom I often worked closely; he had little choice in the matter, since this unusual condition of his tenancy was explained in advance.)

Thus prepared as well as I could be, and with plenty of chicks laid down in the freezer, I collected my new owl from Water Farm and successfully sneaked her up to my flat without attracting attention. When we began our experiment in living together at the end of May 1978 we quickly settled into a vastly more satisfactory relationship than I had ever known with Wellington.

◎◎

"MANNING" MUMBLE WAS never a problem—she was tame from the first moment I met her. She explored the flat with interest, on foot or by means of long, hopping jumps. When I needed to put her in the balcony cage in the morning before I left for work, it was usually no trouble getting her into the cardboard box so that I could carry her out there. She seemed to approve of her accommodation, and usually disappeared into her hutch straight away before I had negotiated my way out again through the Double-Reciprocating Owl Valve. When I returned in the evening and went out to fetch her, I had to wait for her to go through a rather lengthier waking-up routine before she was ready to be sociable and allowed me to catch her, but she was perfectly happy as soon as she was released into the living-room.

Since there was no need to use feeding as a training aid, I would give her her chicks last thing at night—this seemed the obvious thing to do with an owl. As with Wellington, I always gave a particular whistle when I had her supper ready to feed her and at no other time, and, unlike Wellington, she got the idea within two or three nights. Sometimes I didn't even have to whistle—she learned to recognize the rustle of a plastic bag when I opened the fridge, and came without prompting. I would throw her chicks into the opened night cage in the kitchen; when she had hopped in after them I closed the cage, and usually turned the light off and left her to it. (Some nights she seemed reluctant to be shut in, so when she had finished the messy business of eating I would reopen the cage and leave her loose during the night.)

While waiting for me to get up and let her out in the morning after nights spent in the kitchen cage, she would amuse herself by tearing up her newspaper "bedroom carpet" and dropping tiny shreds of it out of the cage until they formed a drift on the linoleum floor below—sweeping these up, and replacing her cage lining, became an almost daily chore. Ripping things up seemed to be her favourite game, and since her curiosity was insatiable I soon learned not to leave anything vulnerable lying around.

When she was loose in the flat and I wanted her to come to me for any other reason I would catch her eye and tap a finger on my shoulder, and she would hop up at once—or, more often, down. As soon as she moved in she was able to jump from any convenient bit of furniture up to the top of the open living-room door. From there she had an un-obstructed view of the whole room and of the distant world outside the window-wall, and it immediately became her favourite perch. At weekends, when I let her stay loose around the flat all day, she would doze comfortably on the door top for hours at a time. (When she awoke from a day-time snooze she often gave a little whistling sneeze— "*snit!*"—followed by a shake of her head and two or three ruminative beak-clacks. If I picked her up while she was still a bit dozy she stepped back onto my hand trustfully but

At this age she was still sociable, so a friend was able to take this
photo of us in the living-room one summer evening. The big
spider-plant would not long survive Mumble's learning to fly.

with a slightly careful gait, like a drunk concentrating on
negotiating a flight of steps.)

From her first arrival in the flat Mumble pursued the
natural "branching" behaviour of a fledgeling, manifested
particularly in her case by a fascination with exploring all
cupboards, corners, and crannies. If I happened to leave a
cardboard box or an empty carrierbag lying round, she was
into it like a venturesome kitten; sometimes she would stay
inside it for quite a while, and I occasionally found her actually lying down inside, flat on her front like a brooding
chicken, with her head cocked back. If I left the sliding
doors of the long wardrobe in my hallway open even a
crack, I would soon hear a croon or—after the first couple

of weeks—a fluting "war-whoop," quiet but insistent, echoing from deep inside it: "*w-o-o-o . . . w-o-o-o . . . w-o-o-o . . .*" (It reminded me of the noise we made as children by flipping a hand over our mouths when playing "cowboys and Indians.") She would clamber along the wardrobe in the dark from shoulder to shoulder of the coats and jackets hanging there, until she reached the deepest corner. I wondered if the accompanying whoops had anything to do with some instinct to ask if a dark hole was already occupied by somebody else. She did not seem to be seeking refuge from anything, but was simply enjoying active exploration. (Although some owl species are known for using underground burrows, I've never come across any reference to tawnies doing this.)

One day, when I started to worry about not being able to find her in any of her usual haunts, a noticeably muffled call led me into the kitchen and down to my knees to peer under the table fixed to one wall. I had completely forgotten that, hidden under it, a hole was cut in the plasterboard wall to give access to the water stopcock. Luckily it was too small to accommodate even Mumble's remarkably compressible body, but she had stuck her head into it and was warbling monotonously into the thickness of the wall. She kept this up for several minutes, bobbing and twisting her head as she whooped into the darkness and seemed to listen for a response. I never could decide if she could hear the enticing scuttle of vermin in there, or if it was just another example of her general hole-fetish.

DIARY: 25 JULY 1978 (C. 3 MONTHS OLD)

Her tameness is apparently unaffected by growing, though she starts, and sometimes flies to my shoulder, at loud noises—a gunshot on TV, say, or loud music when I change radio channels (she's fine with Thomas Tallis or Linda Ronstadt, but is less keen on Stravinsky or the Stones). She has even behaved beautifully when visitors came round to see her. [My friend] Bella brought [her daughters] over for tea the other afternoon, and they were charmed by her. She let them sit close to her on the sofa and stroke her fluffy chest; the girls made satisfactory "*aaah . . .*" noises, and it was as much as I could do to stop them actually picking her up in their arms to give her a hug.

She is growing at an astonishing rate, but still looks like a stuffed toy—though a badly frayed one, with lots of feathers pushing through diminishing patches of down (I find tiny bits of shed down everywhere). Her colouring is more marked now than a couple of weeks ago. The facial disc is sharp, with a hedge of tiny dark brown feathers growing backwards from its edges, and a dart of dark brown edged with white growing downwards between her eyes. The crown of her head is still covered with pale grey fluff, but this "stares" apart when she bends her head, and is retreating backwards—soon it will be limited to the back of her neck. Otherwise, feathers are pushing through and joining up all over her: first came the wings and tail, then the back, then the breast and the edges of her face, and now the head. Since she was about ten weeks old her breast has been "ermine"—downy, creamy-white feathers with dark brown central streaks—but there are still thick, fluffy grey petticoats low down on her body.

During the two months she has been living here her flying skills have improved steadily, from accurate, quite long-range powered jumps to deliberate flights from Point A to Point B. From her first day she could already make a single daring leap from the back of this armchair, out the living-room door, across the end of the hallway, through the open kitchen door opposite, and onto the kitchen table—a good 12 feet. By about her fifth week here she was completing fairly complex circuits-and-bumps around several points, and was even managing to hover briefly in mid-air. But her landings are still lousy—very hard, barely controlled crash-landings.

When I came into the living-room with this notebook she was sitting calmly on the back of my armchair. When I sat down she pranced up beside my head and started pecking at my hair. Then she hopped down onto the coffee table beside me, and from there to the right arm of the chair close by my elbow. She is sitting there now, 6 inches from my moving pen, apparently fascinated by it. With quiet care, she lifts her right foot and gently but firmly stabs the wrist of my sweater, before bending to give it a chew, but her heart isn't in it. Her head bobs and weaves slowly, eyes following the pen.

Every minute or so, when my writing hand reaches the end of a line nearest to her, she has a gentle chew at my hand, the pen, or the edge of the page; then she seems to lose conviction, making vague claw-passes in mid-air like a punchy old boxer. Occasionally she jumps to my wrist and rides it for a while, which makes writing tricky. Sometimes she loses interest and rotates her head on its "ball mount," staring straight upwards, or over at the windows, but mostly she remains intent on the moving pen. She jumps from my wrist to the coffee table, has a quick fluff-up, stands briefly on one leg, then hops back to my raised knee; when the edge of the notebook nudges her she complains, with a soft "*kew . . . kew . . .*"

Her vocabulary seems to be changing. For the first couple of weeks it was a single-note squeak or croon; she apparently speaks through her nostrils, since her beak remains nearly or actually closed while her throat-feathers swell. The only other sound she made was a rapid chittering when annoyed or frustrated—sometimes loud, emphatic, and accompanied by beak-clacks, and sometimes a low, evil muttering. But over the past couple of months her range has developed, to a broken two-note sound like a rusty gate creaking, and very occasionally she has tried out a long, wavering whistle, as if she is practising for a full hoot.

At the weekend I'm going down to [my sister's home at] Petersfield, and I'm going to take Mumble with me—she'll be comfortable enough in the summerhouse over night.

30 JULY

The weekend at [my sister] Val's went well. I had the rear seats of the car folded down, and rigged a plastic sheet over the backs of the front seats and right back over the rear cargo area. (I also decided to wear the ratty old field jacket while driving—I don't want owl-crap down the back of my shoulder when I'm wearing anything halfway decent.) She was quite happy in her temporary overnight quarters, and when I brought her indoors she let the family handle her. When she sat on [my nephew] Graham's shoulder she was calmer than he was—which was fair enough, since it was his first time.

During the drive back on Sunday night she rode the back of the passenger seat for a while, and sometimes went stomping back over the plastic sheeting so she could look out the rear window—she did this when the radio boomed out the choral opening to "Zadok the Priest." But she spent most of the trip on the back of the driving seat just behind my neck, looking sideways or peering forwards around my head—silent, calm, and balanced. Occasionally I felt a warm, companionable little nudge, as she turned and pushed her head against my cheek or ear like a sleepy child. The incredible softness of it brought back a sudden sensory memory, of my five-year-old self sneaking into the bathroom to experiment with one of Dad's old badger-hair shaving brushes.

When we got home and into the underground car-park she had to go back into the cardboard box for the trip up in the lift, with me holding the open side of the box against my chest. (I'm still using a box to smuggle her in and out, because it's less of a give-away than a basket in case I run into somebody in the lift.) This time Mumble absolutely hated being put back into the box, and all the way up she clung flat to my chest with half-open wings, like a moth—though luckily she never hoots on these occasions. As soon as we were inside the flat and I eased my grip, she squeezed out through the widening gap, whiskery face first, and flew to her door-top perch, where she stood shaking out her feathers and giving me dirty looks.

◉◉

WHEN MUMBLE WAS sitting calmly on the back of my chair, I could turn my face and—seduced by her voluptuous fluffiness—nuzzle her chest, with no reaction beyond a quiet croon and a shifting of her feet; but I didn't try this on those occasions when she showed she was in the mood for some rough romping.

Typically, she would walk across the floor and jump up to my raised foot, then climb up the incline of my shin with the aid of half-beating wings. Pausing on my knee, she would then pounce into the crook of my bent arm, excavating it like a terrier, as if pretending that there was a mouse down among the side-cushions. If my sleeve was rolled up she used to peck her way gently up the hairs on my arm, as if she was eating corn-on-the-cob. She might then jump up to the middle of my chest, "footing" it vigorously and making wheezing noises, with her eyes wide and excited (she seemed to find it irritating when I was wearing a sweater, and would lift her feet fastidiously as if trying to get her claws out of sticky tar). These mock-hunts usually developed into pecking at my beard and moustache; if I got bored with this I would catch her beak gently between my knuckles, but there was clearly an inhibition at work

Mumble giving my ring-finger what can only
be described as a close appraisal. She is
gripping it gently between her two forward
toes, but the needle-sharp talons are
not to be ignored.

here—she always pulled her punches, and even in her rare
"berserker" moods she never threatened my face.

She could often be diverted by the rattle of my dropping
a pencil onto the marble coffee table at my elbow. She would
transfer all her attention to kicking it, picking it up, gnaw-
ing it, dropping it, then starting all over again, and sometimes she carried it off to continue the game elsewhere.
I rigged up a swinging pencil for her, hanging crossways
at the end of a leather thong from the "gallows" of her

tray-perch, low down so that she could stand in the tray and reach it with a raised foot. She seemed to like this toy—she sat holding it, biting it, knocking it away and grabbing it as it swung back again.

I noticed that after a period of any kind of exciting activity she seemed to get tired quite easily, and went off for a brief snooze to recharge her batteries. Sometimes this meant sleeping one-legged on her door top, but at other times she would seek out a favoured front corner of the shelf inside the top of the hallway wardrobe, and actually fold her legs up under her chest and lie down flat on her front, like a chicken.

◎◎

DESPITE MUMBLE'S OCCASIONAL over-excitable teenage outbursts, she was much more often docile, and in this mood she definitely enjoyed being "preened." When she was sitting close to me I would stroke the top of her head gently with a finger. At my first touch she would twist and "gimbal" her head around a bit, and after the first few strokes she went quiet and dopey, eyes slitted, crouching low and puffing out her feathers as she enjoyed the sensation.

I was not expecting any deliberate gestures of companionship on Mumble's part, and it came as a pleasant surprise when she started initiating such preening sessions herself. These most often happened when she saw me for the first time in the morning. If she had been in the night cage, then when I walked into the kitchen she jumped down to her newspaper floor, and warbled a quiet series of long, liquid "w-o-o-o . . . w-o-o-o . . ." whoops, with her head shoved low down into a corner of the cage. When I opened the cage door she hopped up onto the doorstep perch and sat blinking up at me sleepily; I would blink back, and when I leaned down to her she turned her face up to me, closed her eyes, and nibbled gently at my beard while I nuzzled her. After a few moments of these mutual greetings she would hop rather deliberately up onto my shoulder, and ride me around the flat until she decided which perch she fancied.

I didn't always shut her up at night, and when I emerged from the bedroom after nights during which I had left her loose in the flat she would be sitting in the semi-darkness on her favourite night perch—the top of the half-open bathroom door. When I went in and turned on the light she jumped over to the shower rail, then to the top of the wall cabinet, where she sat bobbing and weaving, and pecking absentmindedly at my hand when I rummaged blindly under her skirts for the shaving soap. When I had washed my neck and started lathering it, she hopped to my shoulder without prompting, and turned to face the mirror. She never made the classic animal's response to a mirror—trying to find "the other owl behind it"—which was intriguing, because she had only ever caught the briefest glimpses of other owls when she had been a hatchling and during her couple of days at Water Farm. (Recognizing themselves in a mirror argues a degree of self-consciousness, which is generally considered to be a sign of high intelligence in animals.) Sometimes she started clicking her beak through the side of my hair and beard—so I did it back, pecking her breast feathers with stiffened lips, and trying to imitate her "machine-gunning" sound with my tongue.

When I got into the shower after completing my shave

she seldom showed any interest, for which I was grateful. On the first couple of mornings she had sat on the shower rail and watched me beadily, making me wonder uneasily whether she might try to explore me more directly when I turned the water off. For some reason, it was on occasions when I had no clothes on that my eyes were most drawn to her talons.

◎◎

SHE NORMALLY WANDERED off when I turned the shower on, but every day without fail, soon after the water noise stopped—and usually while I was sitting on the side of the bath, towelling my feet—I would hear a questioning croon, and her face appeared again, wide-eyed with interest, around the bottom of the bathroom door. For some unknown reason she had decided that this was the perfect opportunity for some heavy necking.

She advanced across the floor in two or three great leaps, with half-open wings fanning slowly, until the last jump carried her up to my knee (while I tried to cover my modesty—and, more to the point, my vulnerability—with the towel). She shuffled closer along my thigh until she was beneath my chin, then arranged herself neatly, gave another croon or two, and lifted her face towards mine with beak half open.

When I lowered my face to hers she briefly pecked and preened at my beard. When I bent lower and nuzzled her head with my nose, she went tense with ecstasy, flicking the nictitating membranes—the inner "second eyelids"—across her eyes while she twisted and rubbed her head against the gentle pressure. Steadily, she sank further down on her un-dercarriage and fluffed out her body feathers, retracting her head into her "shawl" until she looked rather like a single fluffy ball with a set of features on its top surface. Her head and neck smelt delicious—clean, warm, woolly, and sort of . . . *biscuity*.

If I stopped nuzzling her for even a moment she squeaked insistently, shoving her face upwards. She loved it when I rubbed the close triangle of short feathers immediately above her beak and between her eyes. While I did this she would sometimes actually revolve slowly on the spot until her head was facing away from me, and we were "rubbing noses" while gazing into each other's faces upside down. After a few moments of this she would start twisting her head again, presumably bringing new itchy areas into contact—often, the feathered flaps of the big ear-trenches hidden behind the edges of her facial disc.

When I finally got fed up with the crick in my spine and regretfully insisted on bringing the session to a close, Mum-ble would complain briefly, but would then sit for a while in a sort of afterglow. Then she shook herself vigorously, and jumped up to my shoulder to look around brightly for new amusements—this was my permission to move.

◎◎

MUMBLE'S USUAL EVENING patrol routine involved short flights from chair backs to door tops, to her tray-perch, then down to the living-room floor to walk along the foot of the window-wall, then up to a bookshelf, then out into the darkened hallway to fly to the telephone table, then back to

Mumble at about eleven weeks old, in July
1978; her wing and tail feathers are fully grown
and her front is starting to show the "ermine"
effect, but there's still a lot of fluff on her body
and head.

my chair. She occasionally took a short stroll on a glass-
topped bookshelf, clicking along with a gingerly gait as if
walking on ice. Infrequently, a hollow echo of this same
"gunfighter with spurs" sound betrayed the fact that she had
flown down into the empty bath and was stomping up and
down there—perhaps in search of anything intriguing that
might have crawled up through the plughole?

When I wasn't watching her, there might be a sudden crash as she leapt from some perch straight down onto the plastic sheet covering the sofa. She would give it three or four quick, killing kicks, then move around on the unstable surface of the cushions with wings nearly spread and "mantled" downwards and forwards, in the pose that raptors adopt to cover and defend a kill. Then, satisfied with her victory, she would fly up to one of her perches and fluff her feathers, flick her wings back neatly like coat-tails, and give a final little shuffle before settling down and tucking one foot "in her pocket."

I never tired of watching her floor-walking, largely because she found it so interesting herself. Before she jumped down there from a perch she considered the drop zone carefully, head on one side, as if making calculations before she committed herself to a plan. Once down, she would stroll across to sit at the foot of the window-wall in her "cottage loaf" pose—bum planted squarely on the carpet, legs retracted so that only the tips of her talons showed under her skirts—and gaze around alertly with wide eyes. Often she seemed to spot some invisible real or imaginary prey a few inches in front of her toes—which was odd, because I knew that her short-range vision was bad. Nevertheless, she would stare fixedly at one spot on the carpet before jumping up to full leg stretch, pouncing with murderous violence, and "killing" it. She might repeat this game for a full minute at a time.

If something outside the window-wall caught her attention she sprang into movement, extending her legs like a chicken standing up and making a bobbing run along beside the dark glass, balancing herself with wings billowing half-open like a pantomime villain's cloak. This "sinister stalking" effect—reminiscent of the cartoon cat Sylvester trying to sneak up on the canary Tweety Pie—was particularly amusing when I saw her dashing from behind one bit of cover to another, stopping, then making the next short rush. Naturally, she liked "tunneling" into dark corners under low bits of furniture, but also stomping around boisterously underneath things that had a taller clearance. Predictably, the newspapers on various parts of the floor would often suffer catastrophically under her joyous attacks.

◎◎

I DID TRY giving her a Ping-Pong ball to play with, but after kicking it around listlessly a couple of times she completely lost interest. Since she couldn't get a claw into its hard surface, she didn't seem to think it was worth chasing— crumpled-up balls of newspaper were much more fun. She was also delighted with a gift from one of my friends: a light, plush ball of the kind that mums hang from the tops of prams and cribs. She killed this within seconds, bore it up to her door top (with some balance difficulties during the flight—it was nearly the size of a tennis ball), and proceeded to disembowel it, expertly. Within fifteen minutes the floor was littered far and wide with its stuffing of synthetic fluff. I didn't want her to swallow any of this dubious stuff, which was no doubt made from some petrochemical by-product, and it was clear that if I got her any more of these toys they would last only minutes, so I didn't repeat that particular treat.

For some unfathomable reason, she seemed to be fascinated by my feet. If she was down on the floor she stalked them silently when I walked past, making me nervous that I might step on her. When she was up on her door top she often watched them moving across the floor with intent focus, calculating their range, course, and speed before curling her talons over the edge of her perch, dropping her head between them to keep her eyes centred on the target, and then launching herself unerringly.

I soon began to wonder if the real objective of these attacks was my shoelaces, for which she revealed an un-quenchable lust. Sometimes she would stroll innocently across the living-room floor to sit on the carpet beside my chair, her quiet little head apparently watching the TV screen, but before long I would feel the tap of her landing on one of my crossed feet. Settling herself firmly, she bowed her head and began nibbling and tugging at my laces. Her sharp, hooked bill was shockingly destructive, and when I finally could no longer be bothered to keep shooing her off she was capable in minutes of reducing a woven lace to a drift of separated, broken threads on the carpet below my feet. I soon had to replace all my laces with leather thongs; the discarded worm-corpses of the woven ones, discovered in wastepaper baskets, were among her favourite playthings.

DIARY: 11 AUGUST 1978 (C. 3.5 MONTHS OLD)

Tonight was a first—it seems trivial, but it's yet another pleasing contrast to my experience with Wel-lington. Normally, when I go into the balcony cage to fetch her indoors after coming home in the evening, I wait for her to go through her routine of "whooping" in the corner of her hutch before emerging, and then sitting on her doorstep perch for a couple of minutes while she gets herself organized. Meanwhile, I stand with the cardboard box under my left arm, open end towards her. When she seems to be settled, I reach out and slide my right hand up behind her legs so that she steps backwards onto it; then I guide hand and owl together into the box, simultaneously swivelling it round against my chest to leave only a narrow crack from which to extract my hand. I then try to get back indoors before her frustration provokes painful efforts to dig her way out through my chest. This may be accompanied by chittering, and a furious, whiskery little face trying to squeeze its way into view around the edge of the box.

Tonight, as I stood waiting for her to finish blinking, yawning, stretching, crapping, shaking her feathers, and generally going through her waking-up routine, she looked at the box calmly, measured the distance, and jumped right into it. Then she turned round so that she would be facing the right way "when the lift door opened," and stayed still and quiet while I went through the performance of extracting us both through the Double-Reciprocating Owl Valve and into the living-room.

Her vocabulary is continuing to expand rapidly. Her range of sounds still includes the original cheeps, croons, and squeaks, the more recent creaking whistles, and the habitual, tremulous "Indian whoops" that are always directed into some enclosed place. "Beak-clacking" is a sign of annoyance, but apparently it's not always aimed at other people. She also does it sometimes when simply waking from a doze, alternating short bursts of clacking with sneezes. The first time I managed to watch her closely while she did it I discovered that she wasn't actually snapping her beak open and shut at all—the beak stays half-open. I suspect she must be snapping her tongue from its resting-place pressed into the floor of her beak, and the noise is the "clop" of broken suction—just like the sound that we can make with our tongue and palate. (So she's not doing anything so crude as banging her mandibles together—she's actually speaking Xhosa.)

Most notably, she now has a proper five-part hoot: "*Hooo!* . . . (three to five seconds' pause) . . . hoo, hoo-hoo *HOOOO!*" in a descending tremolo. [I have since read that a bird species' characteristic call or song is partly a genetic inheritance and partly learned by imitation, so Mumble must to some degree have copied her hooting from wild owls that she could hear calling in the distance, though I couldn't.] She also has a sharply bitten-off "kee-*wikk*!" which sounds to my vulgar human ears like a rude Anglo-Saxon expletive. I heard her doing several of these in her night cage recently, about thirty minutes after lights out. Sometimes she also joins briefly in the dawn chorus at about 5:00 a.m.—just half a dozen hoots, then silence.

◎◎

FROM OUR FIRST days together, Mumble's method of launching herself from the top edge of the living-room door when I walked past was always impressive. Without stretching from her ball-of-fluff pose, she would simply lean forwards and roll confidently off into space, hardly opening her wings as she dropped to my shoulder, and landing with a little squeak. Given my own brief and painfully incompetent flirtation with parachuting two years previously, I was deeply envious of both her equipment and her technique. Naturally, as the weeks passed I sought every opportunity to watch her practising her flying. This learning process took time, and since she lacked the luxury of an instructor with dual controls to keep her out of trouble until she was ready to go solo, the weeks while she still had her "L-plates up" were not without incident.

Take-offs and point-to-point flights came easily. From a solid perch, she sprang into the air by the power of her flexed legs, gave one downstroke of her spreading wings, and immediately achieved flying speed. The large size of her wings compared with her weight gave her a "light wing-loading," and thus effortlessly buoyant flight. (Owls have only about one-third the wing-loading of, say, a duck, which has to flap its wings frantically while charging down a long runway for take-off.) When she was accelerating, her individual wing movements were too fast for my eyes to isolate, but stop-frame photography of tawnies has shown me that the long, finger-like primary feathers at the wingtips curl forwards at the ends of the up- and down-strokes, making crescent patterns in the air. The broad, yard-long spread of her wings made her body look like a small, perfectly horizontal nacelle, with her head streamlined face-forwards into the shape of it by her ruff of neck feathers. Her tail was furled into a single "spike" when she first took off, but then spread out into a rounded fan. Her undercarriage was retracted, with lower legs bent up parallel to her belly, feet almost invisible against the feathers, and claws neatly folded in under her toes.

Nevertheless, despite Mumble's elegant mastery of most phases of flight by her first autumn, it was a while longer before most of her landings improved beyond the frankly lamentable. She was fine when she flew up at a perch at an angle from below, but controlling any kind of downward swoop seemed beyond her. The problem was clearly the difficulty of co-ordinating the transition from horizontal to vertical movement (exactly like a novice RAF pilot mastering the tricky technique of landing a vectored-thrust Hawker Harrier "Jump Jet"). Until she got the hang of this, she usually flew full tilt at the objective—her tray-perch, the back of a chair, a table lamp—and smashed into it, sometimes actually rocking it with the impact and giving a little winded squeak.

Her incompetence was most obvious when she forgot (as she often did) her previous experiences of "ice landings" on the long marble coffee table. She made her final approach far too fast and at far too shallow an angle, and when she touched down she simply skidded across it with her feet flailing vainly for traction, her wings beating wildly, and her tail fanning out awkwardly in all directions. Invariably she would disappear off the far side in an ungainly cartwheeling of feathers, like some broken ornithopter in a jerky early-1900s newsreel.

One day she added a refinement to this routine, attempting a lengthways landing while a roll of paper towels was lying on the near end of the "flight deck." Apparently expecting this to be solidly fixed, she smacked straight into it, talons-first. Naturally, under the impact of about a pound of fast-moving owl it began to roll along the table, unreeling as it went. Mumble found herself frantically flapping and back-pedalling on top of it, riding the ever-diminishing cylinder like a lumberjack on a rolling log until both of them fell off the end of the table. She seemed to find my helpless laughter irritating (one can take umbrage so much more convincingly when one has a lot of feathers).

◉◉

IN TIME, OF course, she mastered the landings too, learning by trial and error to pick the moment to "flare" her wings. As she approached the selected landing-spot her body began to swing down from the horizontal towards the vertical, rotating around the axis of her shoulder joints. Simultaneously, she began to bend her wings up at the "elbows and wrists" into shallow L-shapes, and to flare them—tilting the leading edges upwards to increase the angle of the "aerofoil." Meanwhile, her tail spread downwards in a near-vertical fan. The tilting slowed the airflow over the top surface of her wings and so reduced the lift beneath them, while the presentation of the underwing surfaces and tail-fan acted as air brakes, cutting her forward speed. At this point, as the airflow above the wing broke up, an aircraft might stall and fall out of the sky, but at the critical moment Mumble extended the alula feathers part-way along the leading edge of her wings, which both speeded up the airflow and increased the wing area—the principle behind the leading-edge slats that airline passengers see and hear thumping out during final approach.

Meanwhile, she had been extending her legs and swinging them forwards in front of her near-vertical belly, spreading the taloned toes to give a wide reach. Lift was killed at the moment of touchdown, but not always forward speed—usually she sank down as surely as a helicopter, but if she was excited she still made a fighter pilot's fast, flashy "combat landing." Either way, the first part of her to make contact with a perch was the tough, nubbly soles of her feet, at which point the talons instantly snapped closed to grip.

For a "pick-up of prey" rather than a landing, she glided in on horizontal wings with her legs extended. She kept her speed down but did not kill all the lift, bringing her feet forwards at the last second to make her hit, before another wingbeat lifted her away again.

ONE THING THAT I had not expected was the discovery that Mumble sometimes used her spread, lowered wings to support part of her weight when "grounded." I first noticed this when she was playing boisterous hunting games on the plastic-covered sofa, but a more dramatic example was an attempt one evening to land on a four-string set of clothes-lines stretched above my bath. Attracted by furious squeaking, I found her trying to deal with the trampoline effect caused by the sudden arrival of her weight on the separate stretchy cords. She was bouncing around, gripping two of the strings with her feet and trying not to do the splits as her weight pushed them apart. Wobbling and lurching, she extended her wings out and down in the "mantling" pose like a spread cloak, resting them on the cords to stabilize her balance. (The spectacle reminded me of a David Atten-borough natural history programme on TV, featuring a primitive Venezuelan bird that retains a vestigial claw on the alula in the leading edge of the wing—a survival from the reptilian past—which the young bird uses to help it clamber around the treetop nest.) This confident use of her wings as general-purpose limbs surprised me; I had ex-pected her to be completely protective of them.

◉◉

DURING THE SUMMER of 1978 I broke my left ankle, re-sulting in six weeks in a plaster cast from knee to toes. (The event was supremely unheroic: I slipped on wet grass while knocking a tennis-ball around with a friend's little daughter—who, incidentally, subsequently grew up to illus-trate this book.) As anyone who has tried it will know, al-though the initial pain wears off in a few days, and you soon learn to get up a reasonable speed over level ground by swinging between the crutches, the inconvenience of trying to follow the simplest domestic routine while wearing a heavy cast is infuriating. I also felt ridiculous in public, with one leg of my oldest pair of jeans hacked into rags that were held together with safety pins around an increasingly grubby cast. Since showering was impossible, and taking a bath ridiculously difficult, a certain feeling of shamefaced squalor also crept over me as time went by.

The cast made the chore of getting through the Double-Reciprocating Owl Valve to carry Mumble in and out of the flat complicated and time-consuming, and I tended to leave her free indoors as much as I could. One evening during September I went out to the nearby cinema for a couple of hours, leaving Mumble loose to roam the hallway–bathroom–kitchen–living-room circuit. When I got home at about 10:30 p.m., she was gone.

I lumbered frantically around the flat, calling her and checking every cupboard and corner. To my fury, I found that I had carelessly left a small upper window in the kitchen open a crack too wide. Mumble had never shown any interest in it before, and I had grown complacent. I began to face the fact that my owl was gone, and—worse— that she would have very little chance of surviving for long, out there amid the concrete, the traffic, and the other humans whom she had never learned to avoid. It was miles to the nearest stretch of woodland, and anyway she had never been shown how to locate and catch edible prey. It was not even the mating season, when I could have fantasized that she would meet up with a male attracted by her call and might conceivably learn from him how to hunt. I cursed my careless stupidity, and I realized immediately how miserably I would miss her presence around the flat on otherwise solitary evenings. (It is my character to respond to probably bad news by imagining the very worst that could happen, and then steeling myself to cope with it; that way, anything less than the worst comes as a relief.)

Rather hopelessly, I turned on all the lights and opened all the windows, just in case she was close enough to find her way back in if she wanted to. But she had never before seen the apartment building from the outside, and had no way to orientate herself—how was she to recognize one lighted box among sixty-four identical boxes? I peered out and down and sideways, at the narrow concrete ledges that ran around the building at each floor level, but there was no sign of her. And even if I had spotted her, what could I do? I have no head for heights; I couldn't have climbed out and along an inches-wide ledge even if I had been fit, let alone with a leg in plaster.

I hung out of the windows, feeling futile, but giving Mumble's "suppertime" whistle until my lips cramped. When I finally gave up and went wretchedly to bed, I left a chick hanging on a string from the catch of an open window, with a light on behind it. I had the crazy hope that when the window was finally the only lit one in the blackened building she might notice it and be attracted inside.

◎◎

AFTER A LONG time I finally slipped into a miserable half-sleep, but at about 2:00 a.m. I came fully awake again: I would *not* accept this without one more try. I tugged my slit jeans on over the cast, slipped into a moccasin and a blanket jacket, and went around the flat peering and whistling out of the windows. The night was cooler now, and windier. The stacks of lower buildings that surrounded the foundation deck of my high-rise were mostly in darkness, and apart from lone cars on the main road, about the only sound I could hear was made by occasional drunks, reeling out of a nightclub-cum-dancehall at street level in a neighbouring building. My attention caught by this, I happened to be looking straight down into the well beside my building when, silhouetted against the faint wash of security lights at ground level, I saw it—a little black broad-winged shape, gliding across.

Exhilarated that I seemed to have a chance at last to do something positive, I grabbed a torch, Mumble's travelling basket, and a bagged chick, and hobbled to the lift (I didn't need the crutches any more, but I still moved like a bad imitation of Long John Silver). Almost as soon as I lurched around the concrete deck to that side of the building and whistled, I heard a croon. I shone the torch upwards, and there she was, on a ledge about thirty feet up the office building next door.

The next solid hour and a half (I swear) were among the most nerve-racking of my life. Grimly, I staggered back and forth around the foot of the building, whistling and waving a dead chick at the night sky. I was constantly worried that any one of a number of scenarios might suddenly develop. I was afraid that Mumble would get bored and fly off into the distance, never to be seen again. I was worried that a neighbour's bedroom window would shoot open, a torch beam would pick me out, and an angry voice would demand to know what that lunatic was doing lurching around at this time of night, waving a dead chicken in the air.

It was always possible that I might have to explain myself to a policeman—I was already trying to ignore some very old-fashioned looks from a night security guard inside the lit window of one of the ground-floor offices. Perhaps my worst fear was that one of the happy drunks leaving the dancehall doors about fifty yards away would decide to come round the corner to relieve himself in the shadows. (I could just imagine the conversation: "Wotcher doin,' mate? Tryin' to catch an owl, eh? 'Ere, Steve—Steve! Bloke 'ere wiv one leg, tryin' to catch 'isself an owl . . . Let's 'elp 'im, shall we?")

Mumble moved about regularly, from ledge to wall to ledge, almost always within sight, never within reach. She spent a long time looking down from her perches and talking to me quietly, especially from the pierced concrete-block wall around a boiler room; only ten feet tall, it was tantalizingly climbable—for a man with two functioning legs. I had to turn off the torch to search for the flash of her white breast against the dark backgrounds, and sometimes when she was out of sight bits of wind-blown litter glimpsed out of the corner of my eye would raise futile hopes. Once or twice she peered down at me intently, dropped her head between her toes, and seemed about to launch herself towards me—then the rattle of a rolling tin can or the rustle of a discarded sheet of newspaper would distract her, and it was all to do again. I became increasingly tired and confused, alternately cursing this infuriating fowl and, in times when I lost sight of her, rehearsing in my mind a post-Mumble life, and realizing glumly how much I would now hate living here without her.

It must have been at around 3:30 a.m. when, stupid with fatigue and disappointment, I sat down on the edge of a concrete planting trough to rest my plastered leg, which was now aching painfully. I had not been able to catch a glimpse of Mumble for some time. At least the drunks had long since dispersed, and the night was silent apart from the wind. I put the basket down beside me, slipped the slimy chick into my pocket, and pulled out a smoke and my lighter. As I bent my head to rasp the wheel of the Zippo, I heard a clicking sound. It was Mumble's claws, on an iron railing beside me.

She stood a yard from my hands, bobbing her head, peering at me, and squeaking. I dropped my cheroot, sneaked the chick out of my pocket, and extended it towards her, muttering abject endearments. She hopped onto the top of the basket, craning for her supper; I withdrew it. When she jumped down level with the basket I let her grab one end of the chick, but didn't let go of it. She chittered with irritation and pulled—I steered the tug-of-war backwards—and at last the moment came when I could shove the whole situation into the basket, let go of the chick, and slam the lid shut.

On the way back up in the lift I was muttering dopey baby-talk, and Mumble was ripping busily into her late supper. Once inside the flat I stumbled around slamming all the windows again before I released her. She behaved as if nothing whatever had happened. Despite being almost nauseous with fatigue and relief, I sat up for another half-hour watching her fondly while she finished her gory meal, had a brief but apparently satisfying grooming session, stretched, crapped, and went happily to her night cage for a long sleep. I couldn't be bothered to struggle out of my clothes before I collapsed on my bed.

◉◉

IN THE AFTERMATH of this bad scare, I was forced to recognize what Mumble had quietly brought into my life, and what I would have lost if that night had turned out differently.

She had moved into the flat at a time when I was chewing over the cold cud of some fairly discouraging insights into my own character and the probable future shape of my life. The timing had been purely coincidental, and I had been completely unaware of any gift that she might bring me apart from a pleasant distraction. I was buying a pet bird, that was all; after the Wellington fiasco the idea that it might lead to any kind of real two-way relationship had seemed remote.

Mumble's arrival had indeed distracted me from glum introspection: I had had to pull myself together and think hard, right now, about practical problems—before she destroyed something valuable, or disappeared into the plumbing, or got injured, or was discovered by the "Feds" and banished. But she had also offered me her unexpected gestures of trust—perhaps even of affection? What's more, she was providing me with a nightly cabaret. Funnily enough, I have never enjoyed human slapstick comedy, and I would rather spend an evening filling in my tax return than watching Charlie Chaplin movies. But I was now discovering that it was quite impossible to sustain a mood of self-centred depression while an indignant ball of feathers was doing squeaking pratfalls all over the place. This was food for thought, and for some optimism.

I decided that it was high time I learned more about my new flatmate, so that I could prepare myself for the ways in which she might behave when she had finished growing up. Mixed marriages can work well, but they aren't straightforward; if Mumble and I were going to have a long-term relationship, it would be a good idea to find out a bit more about her family.

4

◉◉

The Private Life of the Tawny Owl

Tawny owls are found all over the great Eurasian landmass, from mainland Britain (though not in Ireland) right across to western Siberia, China and Korea, down to Iran, the Himalayas, north-west India, and Burma. The north-south limits of their range in the Western world are roughly from central Norway and Sweden down to the Atlas Mountains of Morocco. They must have arrived in Britain at least eight thousand years ago, before the English Channel finally broke through between the North Sea and the Atlantic Ocean and turned us into islanders—tawnies don't have the aerial endurance for long flights over open water.

Adult tawnies measure up to 14–16 inches (35–40 centimetres) long, though their characteristic squatting posture makes them look misleadingly shorter; they weigh between 13 ounces and 1 pound 12 ounces (385–800 grams), and their wingspan varies between about 37 and 41 inches (94–104 centimetres). The female—the hen bird—is usually about 5 per cent longer and up to 25 per cent heavier than the cock. Their coloration provides camouflage in their woodland habitat: barred and mottled shades of either browns or (less commonly) greys, and off-whites—the plumage tends to be grey on tawnies living in northern regions and brown in the more temperate zones. Brown tawnies can be of either a "rufous" chestnut or (like Mumble) more chocolate shades. All of them are highly vocal at night, particularly from October to January. Their familiar hooting has long led to their being regarded as the "generic owl" of Europe, and a recording used as a radio sound-effect instantly sets the scene as "night-time in the countryside."

The tawny's concealing woodland habitat makes its numbers harder to count than those of the Barn Owl, but it is certainly Britain's most numerous owl—indeed, it is our most common bird of prey. Estimates of the Tawny Owl population published over recent decades have varied surprisingly widely, from a low of about 20,000 breeding pairs to a high of 100,000, and one source even put the number of individuals at around 350,000. The honest answer to the question "How many Tawny Owls are there in Britain?" is therefore "Nobody knows—but lots." Although the numbers have seemed to be in slight decline in both Scotland and south-west England, warming of the climate may now be extending their range a little to the north, and the national conservation agencies do not consider the overall population figures to be "of concern." (In Europe as a whole their numbers have been estimated at anything between 900,000 and 2 million, and are believed to be increasing slightly. This is heartening—and perhaps surprising, given the mindless massacres of anything with wings that continue to dishonour the "sportsmen" of many parts of southern Europe.)

◎◎

THE INTRODUCTION IN Britain of chemical pesticides and seed-dressings in the 1950s coincided with an apparent collapse in Barn Owl numbers. Despite legal protection and public campaigns to install nesting-boxes for them, their population is still estimated at only about four thousand breeding pairs. However, while many people simply assume a direct causal link between the use of agricultural chemicals and the decline in Barn Owl numbers through poisoning of their food chain, in fact this remains stubbornly unproven.

The population had actually been dropping steadily for decades before the 1950s, simply because of other aspects of modern farming. Modern metal barns are less hospitable to owls than the old wooden structures, and since—despite their name—about 40 per cent of Barn Owls nest in hollow trees, the loss of 22 million trees to Dutch elm disease has also had a more recent effect. But food is more urgently important to any carnivore than shelter, and it is the Barn Owl's hunting terrain that has suffered most.

The ploughing up of rough pastureland and field edges, the grubbing-up of hedgerows, the great decrease in the number of hay fields since the demise of the plough-horse, and the increase in the numbers of sheep (which crop the grass too close to shelter rodents) have all greatly reduced the habitat for the short-tailed voles that form an important part of the Barn Owl's diet. Since farmers embraced both the combine harvester and modern grain storage, there has also been less food for rats and mice lying around in rickfields and farmyards. The food chain of grain–rodents–owls must have been central to Barn Owl numbers ever since the invention of agriculture. Tellingly, the few studies that have been made of significant numbers of dead owls suggest that chemical poisoning is in fact very rarely the cause of death, while straightforward starvation is far more common. While there are certainly a number of different reasons for variations in owl numbers over time and place, and a complex inter-relationship between these factors, Tawny Owls have clearly shown themselves to be much more adaptable than Barn Owls, and since the 1950s they have survived in far greater numbers.

Ornithologists believe that this success may largely be explained by the fact that tawnies are birds of the woods rather than the fields, are more purely nocturnal, and are more "sedentary" (home-loving) than Barn Owls, and these last two factors are strongly linked. If you live in thick, dark woodland and hunt almost entirely by night, then however good your eyes and ears are you also need to build up an intimate knowledge of your hunting territory. Owls' eyes are wonderfully developed, but they are not a "super-sense": they are superior to ours only in degree, not in nature. If you are a Tawny Owl, you need to be able to fly between the branches even on nights when it is too dark for you to see anything more than a vague difference between shades of black. You need to be able to find your way to favourite hunting perches close to the areas and the runs used regularly by your whole range of potential prey, and to find your way home with a kill—whether to your safe roosting tree or to the nest where your owlets are waiting impatiently. If you are going to exploit all of your home patch's varied possibilities for seasonal "shopping," you need to have a reliable bank of topographical and spatial memories of its whole extent.

Once you have acquired all this vital knowledge, by sometimes perilous trial and error, why would you choose to travel? So, once it has mated, a Tawny Owl may spend its whole adult life in an area sometimes measuring as little as

350 yards on a side, and with the same mate. The Barn Owl does not have such a predictable breeding season as the tawny, and cock birds may spread their favours more widely among several partners. They may maintain a small defended territory around a nest, but to catch their prey Barn Owls have to range widely on the wing over a much larger area of fields. This more opportunistic method of hunting must logically leave them more vulnerable to variations in the number of available prey animals, which can be dramatic and can occur every few years. With a smaller hunting area that it has taken the trouble to get to know like the back of its foot, and with a permanent partner, the Tawny Owl can adapt to such changes much more successfully.

While British tawnies prefer to live in old broad-leaved or mixed woodland, they have also proved resilient in coping with the felling of forest for replacement with commercial conifer plantations, so long as these are broken up with a reasonable number of clearings and rides. (Conifer forests in continental Europe support large numbers of tawnies.) Many British tawnies have also responded to the spread of urbanization by moving successfully into the suburbs, and even into city centres, where they may nest in nooks and crannies on the ledges of tall buildings.

◐◑

FINICKY EATING HABITS are probably the greatest threat to any creature's chances of survival in a changing environment, and if the panda is at one end of the spectrum of adaptability, the Tawny Owl must be close to the other extreme.

A tawny will cheerfully eat almost anything smaller than itself that creeps, wriggles, runs, flies, or swims. Its most frequent prey are small rodents, though the balance of types on the menu varies to a certain extent with the cycle of seasons and vegetation, and with the amount of competition from weasels and stoats. One famous English study of the tawny population in an area of Oxfordshire established that bank voles, wood mice, and a few shrews made up 60 per cent of their diet, and that the whole range of small mammals—including rats, moles, and young rabbits—made up 95 per cent. (Owls usually leave shrews alone; they are too small to be worth much effort, and their defence mechanism is a foul taste.) In the countryside smaller birds make up only 5 to 10 per cent of the tawny's diet, but urban starlings, thrushes, blackbirds, sparrows, pigeons, and sometimes even jays provide the great majority of the prey taken by those pioneer families that have responded to loss of woodland habitat by moving into our cities. (There is even one record of an impressively ambitious tawny taking a mallard duck off a city lake.) Tawnies will also routinely hunt on foot to catch beetles, slugs, and snails, and particularly earthworms. Those living near water eat molluscs and crabs, and some have been filmed wading into the shallows to catch fish—anything from minnows and garden-pond goldfish up to small river-trout.

The Tawny Owl's most characteristic method of hunting is to pick a lookout post in a tree at sunset and to wait patiently, watching and listening for the movement of prey below. When they locate a potential meal they judge its direction and range with great precision, then make a short, direct pounce or glide to hit it with their talons. Unlike

Barn Owls, they do not make hunting patrols—flying back and forth over open ground, watching out for potential prey below them—but they may dive to make opportunistic kills if they happen to spot something that looks tasty during their short woodland flights from perch to perch. Their ability to locate their prey in darkness—at least, on ground that they know intimately—depends in equal measure upon their extraordinarily efficient eyes and ears.

◎◎

ANIMALS WHOSE SURVIVAL depends mostly on seeing what is sneaking up on them, in order to avoid being eaten, generally have their eyes placed in the sides of narrow skulls so as to give the widest possible field of vision. Animals that depend on doing the sneaking up usually have their eyes placed forward in broader skulls; this increases their binocular (i.e., overlapping, and thus three-dimensional) vision, so that they can judge accurately the distance to their potential dinner as well as its direction.

For example, a pigeon has a total visual field of as much as 340 degrees of the 360-degree compass, with a blind spot of only 20 degrees directly behind its head, but only another frontal "pie slice" of about 24 degrees of the total is binocular. When it is looking in a given direction a Tawny Owl's visual field is probably a bit less than half that of a pigeon, but up to 70 degrees of this is binocular vision. You and I also have wide-set eyes, and within our total visual field of around 180 degrees—roughly, from ear to ear—the natural area of binocular vision is about 90 degrees. But although we can also swivel our eyeballs from side to side to cover up to 140 degrees stereoscopically without moving our skulls, we cannot rotate our heads right round to look backwards. Owls can, so they still win hands down.

Despite our folklore, owls cannot see in total darkness, but their perception of "total" darkness is nothing like our own—indeed, under the open sky there is never a complete absence of light. Owls' eyes have evolved to exploit the faintest levels of illumination, and to function well enough even on overcast, moonless nights that appear to us as pitch-black. Various estimates have been published of the superiority of owls' visual sensitivity over that of humans, but defining the exact terms of reference, and devising experiments so as to screen out the many possible "variables," has proved extremely difficult. Extraordinary as it seems, scientists have found that an average nocturnal owl's "absolute visual threshold"—the point below which it cannot even detect the presence of light—is only about 2.2 times lower than ours, and that some exceptional individual humans may even have a lower threshold than some individual owls. (Incidentally, by this measurement a cat performs better than both of us.) Of course, the practical question as far as the owl is concerned is not how superior its absolute visual sensitivity is when compared with ours under laboratory conditions, but how efficiently it can function at the levels of illumination that occur in its natural surroundings.

Experiments have been carried out with Barn Owls in adjustable indoor conditions of artificial darkness, in which the owls flew a course hampered by randomly hanging strips of light card. Mathematical interpretation of the results suggested to the researchers that in these conditions the owls' efficient exploitation of low light might be roughly one hundred times superior to our own, and to that of daytime birds. In the end, however, when illumination had been artificially reduced well below what we would call absolute darkness, the owls collided with the cards, and thereafter they sensibly refused to take any further part in the experiment. There is plentiful evidence that Tawny Owls sometimes collide with branches or even tree trunks on very overcast nights when they are below the thick woodland canopy—the darkest of all natural environments—and comparative studies suggest that they are more prone to collision injuries than daytime birds.

Again, a number of experiments have been carried out in the hope of quantifying the light levels at which an owl can locate, say, a mouse when compared with a human's ability to see it. These produced the suggestion that an owl was some three hundred times better at this trick than a scientist, but the methods used to reach this conclusion seemed to other researchers a bit hit-or-miss. Apart from the failure to ensure sufficiently consistent conditions and to provide "controls," the question of motivation seems to have been ignored: Just how badly did the scientist want to find the mouse, compared with the hungry owl? (And in case the question of body warmth occurs to the ingenious reader, only dead mice were used in these tests; and anyway, a suggestion that owls can detect a source of warmth by infrared vision has since been disproved.)

Despite all these uncertainties, one analogy that has been quoted is that an owl could spot a mouse on a football field that was illuminated by a single candle. This may well be true—but only for a particular owl, hunting a mouse of a particular colour, on a particular football field, in particular conditions of ambient light and weather, and lit by a candle at a particular distance and angle. All in all, it is probably enough for us to know that (a) in darkness owls can make much better use of their eyes than we can; but (b) this has less to do with the absolute relative sensitivity of our eyes than with how accustomed we are to using them in extreme conditions, in conjunction with our other senses. Put simply, it isn't so much that owls have very dramatically better visual equipment than we do; their advantage comes from their using it far more efficiently.

◎◎

BECAUSE EVEN OWLS' ability to see in darkness has its definite limits, hearing is actually at least as important to them as eyesight, and much more important when the darkness is at its deepest.

An owl's highly developed ears are located in two vertical "trenches" in its skull located just behind the edges of the facial disc. The shape of the facial disc gathers sound, because the dense ruff of feathers around the edges forms a dish, and there is some evidence that the consistency of the tiny face feathers also plays a part in this effect. The "pitch" or frequency of sounds is measured in units of kilohertz; human ears (when young—an important qualification) can detect sounds within a wide range of between 2 kHz and 20 kHz, but are most efficient at around 4 kHz. Tawny Owls' hearing is also at its most acute at medium frequencies, between 3 and 6 kHz. It has been estimated that within certain given frequencies an owl's ears are ten times more efficient than ours, although this varies from species to species; for instance, Barn Owls have an optimum frequency of 7 to 8 kHz. Hearing in owls has been estimated to be perhaps three hundred times more acute than that of daytime birds, but—as with their eyesight—its superiority is nothing like that great when compared with either humans or cats. (And, as with eyesight, there are important qualifications to any "headline" figures, to take account of variables such as background noise. Hunting by ear is obviously far more difficult on a windy night in the woods than on a still night in the middle of a mown field.)

An owl's range of aural perception coincides with the sort of noises that a rodent makes when rustling through grass or leaf-litter, and if the potential meal is foolish enough to squeak, then the owl's job becomes very much easier—as the pitch of a sound rises, the owl's ability to pinpoint it in space improves. (Shrews are extremely quarrelsome creatures with high-pitched voices, and their "sound discipline" at night is stupidly lax; we might guess that it is only their vile taste that has stopped owls wiping them out long ago.)

In addition to acute hearing, the more nocturnal species of owls have a highly developed ability to process the information that reaches their ears, filtering out the distracting background clutter and drawing conclusions from the sounds that interest them. Their brains include a sort of "target acquisition computer," which calculates the angle and range of whatever is making that attractive rustling or squeaking noise. This capacity for fine judgement is enhanced by the fact that some species—including Tawny Owls—have asymmetric ears, with one slightly higher than the other and oriented at a slightly different angle, so that even when the head is motionless, a sound reaches each of the two ears after slightly different time delays. In tawnies this asymmetry affects only the soft outer flaps rather than the actual passage through the bone; in some other owl species the difference is more pronounced, and one ear trench is also noticeably larger than the other, giving the skull a slightly uneven shape.

As well as equipping them superbly for night-time hunting, their all-round vision and sophisticated auditory equipment naturally make owls, in their turn, extremely difficult to sneak up on. While it is far from true in other parts of the world, in Britain an able-bodied adult of the larger species of owls has very few natural predators to fear. (There is one report of a Tawny Owl being taken from its daytime roost by a buzzard, at a time when rabbits were particularly scarce in that area; but the owl fought to the death, inflicting such serious damage on its killer that this must presumably be a rare event—predators will always prefer to go for an easy kill.)

∞

BECAUSE TAWNY OWLS are so relatively plentiful in Britain, and so attached to permanent territories, ornithologists have been able to make some quite thorough studies of their way of life. This applies especially to their rearing of families, during a time of the year when, necessarily, they stick close to home and follow repetitive patterns of behaviour.

Tawny cocks and hens first seek mates around the turn of the year, when they are perhaps eight months of age. By this time they have survived the initial shock of their first early winter in the territories that they established during the autumn. Courtship among all territorial carnivores must certainly be a nervous business, since their instinct is to challenge, drive off, or attack all strangers. Tawny Owls show little if any difference in appearance between the genders, but are clearly able to recognize each other's sex and individual identity by means of calls. (Some species of owls, though not tawnies, sing formal duets together at a distance before approaching one another.)

It is often said that cock tawnies give their familiar wavering hoots, which are answered by the hens with sharp "kee-*wikk*" calls. This does happen, but the distinction is not rigid; both cocks and hens make both sounds, though not during the same conversation. Although Mumble might suddenly exclaim "kee-*wikk*!" if she seemed to hear some distant hoot, her initial long-range calls were hoots; the hoot most often seems to be the question and the "kee-*wikk*" the reply, from either sex. (This means that the traditional rendering of the call as "*tu-whit . . . tu-whoo*" merges both question and reply, and puts them the wrong way round—it should be "*tu-whoo . . . tu-whit*.")

If all goes well at what we might call the "e-mail and phone-conversation" stage, the male flies closer to the female and, after a certain amount of eager chasing through the trees, the pair tentatively agree to meet. From that point on body language plays the major part in their exchanges. Once a male and a female Tawny Owl have, as it were, agreed to lay aside their weapons and discuss this situation like adults, they settle on a branch for face-to-face negotiations. Some courting cock birds have been seen bringing the hen a gift of food, to break the ice. The cock utters a variety of quiet grunts, croons, and clucks while sidling forwards and back again, swaying and nodding, alternately raising his wings and puffing out his body feathers, and then sleeking himself down again. He will sometimes be allowed to get close enough to nibble at the hen bird's beak.

With luck, this performance will eventually earn him a softly repeated invitation from the hen; the cock then mounts her from behind, and Nature takes its course. As with many bird species, while the drive to mate is strong, the actual moment seems to be a fairly brief and mechanical affair, distinctly short of ecstasy. To us the process anyway looks frankly impossible, since it involves the alignment of two internal vents hidden by feathers, but it clearly works for them. However, when the climactic moment has passed the pair will then roost together, side by side. Their earlier mutual suspicion is forgotten; pressing closely together, they will spend long periods "snogging"—preening each other's faces, heads, and necks, in what certainly looks like highly pleasurable pair-bonding. (Well, *I* always enjoyed it.)

While we cannot tell whether or not they are invariably faithful, Tawny Owls are monogamous. There are differing opinions, but most ornithologists believe that tawnies mate for life, although the pair do not live together all year round. They often roost separately—notably during the autumn, when their annual brood of owlets has dispersed. Nevertheless, as soon as the pair-bond has formed they establish a joint territory for hunting, and thereafter they will share it for life, working together to expel any competitors. A famous study carried out in the early 1950s over some 1,300 acres in Oxfordshire (led by Britain's doyen of Tawny Owl studies, Dr. H. N. Southern) estimated the pair territories at between 32 and 50 acres—smaller in close woodland, larger in mixed woods and open country. Size is determined by a territory's richness in prey animals: further north in Britain, territories of up to 80 acres have been recorded (and a German study in much less prey-rich conifer forest found pair territories that seemed to cover several hundred acres). The boundaries of this hunting range seem hardly to change from year to year, since both cock and hen defend it fiercely from rival pairs and from hopeful juveniles seeking a first territory in autumn.

In England, mating pairs of tawnies establish a nest in February or March each year, and may well use the same few convenient sites in their territory year after year. As in their eating habits, tawny homesteaders are extremely easy-going. They can't be bothered to build nests themselves, and usually prefer to move into existing holes in tree trunks or stumps—often courtesy of woodpeckers. Alternatively, they may take over the abandoned nests of other large birds such as sparrowhawks, jackdaws, magpies, and crows, or even old squirrel dreys. Unlike Barn Owls, they will not tolerate humans close by, but they may choose to set up home in a suitable corner of a derelict building, and they are perfectly happy to use artificial nesting-boxes (though these have to be of a different design from those provided for Barn Owls). In conifer plantations with plenty of ground-litter, and in extreme areas of their geographical range such as the thinly wooded Scottish highlands, they may even nest on the ground. This casual attitude extends to the interior furnishing; tawnies make no effort to line a nesting-hole with soft materials, since the hatchlings will emerge from the egg well protected by thick, fluffy down.

◎◎

USUALLY THE HEN lays three to five white, spherical eggs over a number of days from about mid-March, and sits on them for just under a month until they hatch; during this incubation period she is fed by the cock bird. Depending upon region, climate, prey availability, and other variables, the staggered hatching of the owlets may be completed during April, or not until June. From the time when the last hatchling breaks out of the egg the cock bird needs to catch at least a couple of dozen prey animals each day, depending upon their size. Each of the owlets needs feeding several times daily; the cock also has to hunt for their mother, who remains in the nest with them, and to keep up his own strength for this exhausting effort. This often forces him to continue hunting from dusk until well into the next morning in order to satisfy his family's endless demands (which is certainly no fun for him, but provides us with our few opportunities to see a Tawny Owl out and about by daylight).

The hatchlings remain in the nest with the hen bird for about three weeks before they achieve the strength and confidence to start creeping outside and moving about. They grow very fast, and their ravenous appetite increases with their size. At this point the demands of feeding the whole family become more than the most dedicated father can handle alone. Thereafter both parents have to take turns in a busy hunting rotation throughout this "fledging" stage, and the fledgelings remain dependent upon their parents for all feeding for up to another twelve weeks. To satisfy their voracious demands, the parents have to provide well over a thousand rodents and small birds during this period. Among daytime raptors the parents will usually break up the prey and feed their chicks with bite-sized chunks. Initially, Tawny Owl mothers will also roughly "butcher" the prey—at least taking the head off, to get the owlets started—but before long the babies' own claws and beaks are strong enough for them to manage on their own when a meal is too big for swallowing whole.

The fledgelings are both fearless and curious; they test their strength and agility at first by exploring the immediate surroundings of the nest, and then by venturing further—a procedure known as "branching." Their parents' child-minding is complicated by the fact that the fledgelings tend to go off on expeditions in different directions; this perhaps has survival value, so that their loud demands for food do not attract a predator that could kill the whole brood at once. While they are still covered with fluffy down, rudimentary wing and tail feathers are appearing through it and growing stronger every day. At first the owlets can only make hopping jumps, but after perhaps a week outside they get more ambitious. Over about two more weeks their attempts develop from "parachute jumps" to short glides, and then to true, flapping (though rather hit-or-miss) flights from point to point.

It is in this period of their growth that people may find "lost" fledgelings on the ground or among bushes. They usually aren't lost at all, but simply exploring; unless they have been injured, they are strong climbers and are probably perfectly capable of getting home by themselves. Unless they are obviously injured or in a dangerous spot, would-be "rescuers" should not ignore their touchingly brave, hissing displays of defiance—even at this age they know what is best for them.

By about seven weeks after leaving the nest, Tawny Owl fledgelings are approaching the same size as their parents. For about another four or five weeks—so for April hatchlings, until, say, mid-July, but for May and June hatchlings, perhaps until August or even into September—the juvenile owls continue to explore their parents' territory. Their instinct to look out for anything moving and to pounce on it seems to be innate. One would expect them to be practising finding and killing prey during this "adolescent" phase, perhaps imitating demonstrations by their parents, and this may be the case. However, while studies are sparse, one published paper did report that during fledging the owlets observed were not actually catching prey for themselves, but continued to rely on their parents for food—for which they called, monotonously, from whatever perch they happened to have reached in their wanderings. (Parents of teenagers will empathize.)

If this surprising report is true, then the youngsters must be confronted with a brutal challenge indeed when—about twelve weeks after they first leave the nest—their parents stop feeding them, and subsequently force them to disperse independently to seek their own territories. At this stage in late summer the family is disbanded; the exhausted parents separate for a few months, although both remain in their shared hunting territory. From that point on the year's brood become competitors, and it is pretty much a case of every owl for itself. (However, there are anecdotal reports of adolescents and mother birds occasionally being seen together subsequently, sitting on adjacent perches and calling to one another in a conversational way.)

◎◎

IN THE WILD, the mortality rate among any one year's tawny owlets varies considerably. Regional differences, and the sheer difficulty of carrying out statistically significant studies in thick woodland, mean that few confident general conclusions have been published. However, one thing that does seem quite clear is that an owlet's early survival chances depend upon the local supply of prey animals, and that this varies widely over cycles of a few years.

By some mechanism that science has not yet explained, owls are among the species that seem able to adjust their brood sizes in advance to match the cycles of prey numbers in their area. In a very "bad vole year," Tawny Owls in that area will not lay eggs at all; in a marginal year clutches will be small, and one or more of the hatchlings will certainly fall victim to competition in the nest; but in good prey years the clutches are larger, and fewer fledgelings die before they become independent. A good food supply also means a shorter fledging period, and studies show that owls that fledge earlier tend to live longer—presumably because they are stronger and more self-reliant when the leaves drop and the weather begins to turn cold.

Although Tawny Owls are noted for being splendidly fierce defenders of their nests and young, and will not hesitate to attack even humans when they perceive a threat (more than one naturalist has lost an eye to a tawny attack), the owlets are naturally vulnerable to predators as soon as they start to explore their surroundings. While they are always in some danger during their clumsy early expeditions, the degree of risk from predators—in Britain, mainly from jackdaws, sparrowhawks, goshawks, buzzards, and foxes—seems, again, to be correlated with the local supply of small prey animals. To some extent the predators' diet overlaps with that of the owls; if rodents are plentiful the predators are less of a threat, but if they are scarce then predators are more likely to catch owl fledgelings.

A Danish ornithologist calculated that in a "bad vole year" 36 per cent of the owlets in his study area—and particularly those that hatched late—died before ever achieving independence from the parents; the overwhelming majority of these fell victim to predators, particularly foxes. A study in a northern English conifer forest in a year of extreme prey scarcity calculated the Tawny Owl fledgeling mortality rate at a staggering 91.7 per cent. However, the Oxfordshire study already mentioned concluded that in a plentiful prey year the mortality rate up to late July may have been only 4 per cent, and was certainly no higher than 16 per cent.

The death rate after the fledgelings leave their parents and start trying to establish their individual territories is probably far higher, and perhaps as high as 60 per cent during their first six months of independence. The crucial early weeks will often demand fighting—or, at least, convincing threat-displays—against rivals (and Tawny Owls seem usually to be fighters rather than bluffers). The loud hooting heard on autumn nights is the sound of parents driving their broods away, and of neighbouring adult tawnies warning these adolescents off their territory. Simultaneously, the juvenile tawnies must learn to feed themselves by trial and error.

If they are lucky or fearless, they will find or win a territory fairly close by—either because a previous adult proprietor has died and left it vacant, or because the local density of adolescents is low and the competitors are less determined. It is noticeable that when tawny "landowners" die (and there is some evidence that when one of a pair dies its mate may not outlive it for long), their vacant territory is not automatically taken over by the next-door neighbours. Tawnies depend for successful hunting not so much on the sheer acreage of their hunting range as on their intimate familiarity with it, so a property lying vacant is more attractive to a juvenile seeking its first territory than to an established pair.

An owl needs to eat about 20 per cent of its own body weight daily. (Think about that when you next step on the bathroom scales; for me, it would mean digesting 34 pounds every day—the weight of seven volumes of the *Encyclopaedia Britannica*.) If juvenile tawnies fail to catch enough prey to keep up their strength for the nightly hunts and confrontations of autumn and early winter, then they will soon starve to death. Even if they learn hunting skills quickly, heavy local competition may force them to search further afield for a permanent territory. (Occasionally, much further afield: the record may be held by a British tawny owlet that was ringed one May in Northumberland, and was recovered that November all of seventy miles away in Dumfriesshire.)

Even for young owls that do acquire a territory, and make

a start on building up the necessary data-bank of local information tree by tree and yard by yard, the first winter brings steadily increasing dangers both from hungry predators and from simple starvation. There is little reliable data for populations of British tawnies, but some statistics from continental Europe make pretty grim reading. In one Swedish study, the death toll among young tawnies was 67 per cent in the first year, and 43 per cent of the survivors in the second year—so of an original one hundred fledgelings, only nineteen were still alive in their third year. Obviously, the first winter must present young owls with the harshest challenges, but it clearly takes more than one year before they have established mastery over a consistently productive territory. (We might guess that the relatively high death rates in the second year may perhaps have something to do with failure to mate, and thus to establish a sufficiently large joint territory early in that year.)

DESPITE THEIR UNJUSTLY sinister image in folklore, there seems to be only one British example of owls being systematically persecuted by humans in the belief that they were competitors for a resource—but that episode was long-lasting, and distressingly recent. From the mid-nineteenth to the mid-twentieth century, all British raptors, daytime and night-time, were routinely massacred as "vermin" by the gamekeepers of shooting estates, in the belief that this was necessary to protect the chicks ("poults") of the game birds that they were rearing in huge numbers for the guns.

The scale of this immense (and, as it turned out, senseless) slaughter began to diminish gradually after all birds of prey were given legal protection in 1954 by the first Protection of Birds Act, which was updated in 1967. Since the 1970s, the belief that Tawny Owls took significant numbers of poults has been proved to be false by irrefutable large-scale studies undertaken jointly by bird-conservation and sporting bodies. Even allowing generous margins for error, these studies found that all birds of prey were responsible for only some 5 per cent of the poults lost to all causes, both violent and accidental, compared with at least 50 per cent taken by ground predators such as foxes, dogs, cats, and mink. Imaginative co-operation between conservation and sporting interests has since reduced this proportion even further.

Sadly, however, the myth has been so persistent that even today some Tawny Owls still fall victim to illegal, indiscriminate, and barbarically cruel pole-traps. Luckily, since tawnies seem virtually never to eat carrion, they are not vulnerable to the poisoned bait that some keepers still spread around, killing significant numbers of other raptors—especially in Scotland. (Again, even if a build-up of agricultural chemicals in the systems of prey animals is indeed any sort of factor in owl mortality, then tawnies are less vulnerable than Barn Owls, because they take their prey mostly from woods and parkland rather than from the cultivated farmland where most chemicals are used.)

In most of lowland Britain the greatest threat mankind seems to pose these days is through owls colliding with overhead cables, wire fences, and—particularly—motor vehicles. Like high-hovering kestrels by day, they find good hunting along road verges, but their low-level night flights put them in vastly greater danger. (One report noted that on a thirty-mile stretch of road in Dorset, seventy-six owls had been killed by traffic in six months—and those were just the corpses that were found actually on or beside the road.) Even so, British studies suggest that our Tawny Owl population has remained fairly stable over recent generations, with an annual mortality rate averaging about 20 per cent among mature birds. The healthy overall numbers certainly show that enough of them breed and rear families successfully to sustain the national population.

Individually, however, a life in the wild is always high-risk; although one ringed tawny was recorded as reaching the truly astonishing age of twenty-one years and five months, their typical life expectancy seems to be only about five years. Depending upon your emotional responses to the natural world, you may or may not consider a wild tawny's life to be nasty and brutish, but it is certainly short. I continue to take comfort from the thought that Mumble never had to take her chances in this merciless lottery.

5

◎◎

Mumble in Her Pride

A T THE BEGINNING of 1979 Mumble was a fully grown, fully feathered owl, and in the months to come both of us would have to get used to some of the implications of that. In the wild, she would by now have acquired—and would be defending—a hunting territory, and would be on the lookout for a mate. I supposed that in our first months together she had come to regard me, the provider of food, as her mother. It was logical to imagine that over the coming months she might discover a mate-shaped hole in her life, and that to some degree she might expect me to fill it. Since I would be unable to do so except in a very limited and platonic sense, I was fairly uneasy about how she might react to my inadequacy. (It was possible, after all, that Hell might turn out to have no fury like an owl scorned.) In the meantime, I would have to keep a close eye on her attitude and behaviour not only towards me, but also towards other people.

DIARY: 8 JANUARY 1979 (C. 9 MONTHS OLD)

Mumble is still reasonably friendly with visitors, and I don't yet see any evidence of strongly expressed territorial feelings. I never have her loose about the flat when I am expecting anybody, of course—the ring on the doorbell might turn out to be the caretaker, and anyway it would be stupid to risk her exploding out the door past some startled visitor, to go zooming around the public passages and crashing into fire-doors.

If I am urged to bring her in from the balcony cage by somebody who has already arrived and has settled in, I always warn them that it might get a bit exciting. Male friends tend to wave aside my caution, out of genuine curiosity mixed with a hint of machismo—and indeed, when I bring her in and let her loose she doesn't seem actively hostile. She may fly to her door-top perch while she checks them out, but quite often she goes off about her own business, leaving the humans to it. However, both I and any visitor may well get the "shooting gallery" treatment if we move about the flat.

She will be sitting quietly in the shadows on the telephone table at the far end of the hallway, not apparently paying any attention to us—but when anybody walks across the near end of the hallway between the living-room and kitchen doors they are liable to find her swooping out of the twilight at high speed, aiming for their feet. Considering the brief window of opportunity provided by the couple of steps across the four-foot-wide passage, her reactions are incredibly fast (I imagine that in the night-time woods a lot of mice and voles must die without ever knowing what hit them). Whether or not the attraction of human feet is always the amusing shoelaces, I can't be sure. If—as usually happens—the walker freezes, for fear of stepping on her, Mumble skid-turns at the last micro-second and lands behind their feet, as if disappointed that they have spoiled the game. Then she flaps briskly back to the "firing line" to await their reappearance from the kitchen so that she can have another shot. [My friend] Will finds this highly amusing, and takes it as a challenge to his agility; other visitors might look a bit hunted, and make excuses to stay in the kitchen until I have stepped out first to draw the fire.

◎◎

IF MUMBLE DECIDED after a while to explore visitors more thoroughly, she could get so close-up and personal that they might have to display a certain amount of fortitude. Deceived by the picture-book image, they usually seemed to expect that an owl would provide more of a passive spectacle than she was prepared to be, and she sometimes treated them with the same rough familiarity that she showed me. She might try to eat their shoelaces, or—more alarmingly—she might play her "hunting for mice" game down beside them on the sofa, with kicking feet and flapping wings. During her first few months in residence, about the most personal she would get with anybody else was occasionally to land on their shoulder and nibble gently at their ears and hair, but now she sometimes landed directly on their heads. Given my trade, I had several old military helmets around the place, and I began to issue these tin hats to visitors if Mumble was on the prowl. They might prevent an accident, and anyway they served as a useful reminder that an owl was a lot more than a cuddly toy.

I can't recall exactly how old Mumble was on the weekend afternoon when my neighbour and old mate Gerry had come in for a drink, and Will also turned up, bringing an American friend of his named Howard. Howard seemed quite relaxed about my eccentric domestic arrangements, and I knew that while he looked like the proverbial mild-mannered bookstore clerk he was in fact a captain in the U.S. Army airborne infantry, highly decorated in Vietnam. Still, he did have a very bald head, and since I didn't have the authority to write him up for yet another Purple Heart I thought it sensible to dish out steel helmets all round.

My three visitors were sitting in a row along the sofa with their wineglasses when Mumble appeared on the top of the living-room door. She gazed down with interest at the three drab-painted metal turtles below her, tilting and bobbing her head. Then she made her choice, and parachuted onto the middle one. Her claws made a scrooping noise as she touched down on the matte sandy finish of Gerry's martial headgear, and shifted to get her balance.

The conversation tailed off, and three heads turned to watch Gerry. His smile hardly wavered, but his eyeballs rolled up towards the clicking sound on his personal metal roof, and he instinctively leaned his head back a little. The helmet he was wearing happened to be the handsome 1918 Swiss Army model shaped rather like a medieval sallet, with a deep front sloping down to the brim at a rather shallow angle. Naturally, when the helmet moved Mumble stepped forwards to compensate for its backwards tilt, which had brought the front of the helmet almost horizontal under her feet. *Click—clickety, click* . . . Gerry sat still and silent, his eyes straining upwards.

Slowly and deliberately, Mumble shuffled forwards and leaned over the edge. First Gerry saw the gleaming tips of four razor-sharp talons curling round under the front brim of the helmet. Then, between her parted feet, her upside-

An air-landing assault on something rippable affords
a rare view of Mumble's legs at full stretch.

down face appeared, peering back at his. As she held this
pose—intrigued, and considering her next move—Gerry's
expression became indescribable, and the rest of us burst into
helpless laughter. Mumble took off vertically, and I poured
Gerry a refill.

◉◉

JANUARY AND FEBRUARY are the season when Tawny
Owls mate and nest, and despite the fact that we were living
above a busy town centre, Mumble seemed to attract a certain amount of attention from hopeful males during her
first winter:

DIARY: 17 JANUARY 1979

There was a lot of noise from the kitchen cage last night, and again when Mumble woke up this evening. There was a wild owl somewhere outside, hooting repeatedly, and she joined in an exchange of calls. She did not seem to be replying to a hoot with a "kee-*wikk*!" but with a repeat of the hoot. The calls lasted about five seconds, with a couple of beats in the middle: "*Hooo!* (two, three)—hoo, hoo-hoo *HOOOO!*"

During last night I had to get up three times and go into the kitchen to try to hush her, at about midnight, 1:00 a.m., and 2:00 a.m. I'm really worried about the neighbours on that side or in the flats above hearing her and trying to find out where the racket is coming from, but with any luck they will never guess the truth. After all, the idea that the bloke in No. 40 might have an owl living in his kitchen is not the first explanation that would spring to mind.

22 JANUARY

While she was loose around the flat this evening she got into an exchange of opinions with a wild owl. Although I couldn't hear the visitor, she clearly could. She zoomed around from window to window, bouncing on her feet along the floor at the foot of the all-glass wall and then flying to the end windowsill, yelling and peering out intently. Eventually she sat on her barrel on the windowsill [a ceramic keg, originally some sort of advertising prop, which I had picked up in a junk-shop] and hooted so monotonously that in the end I pulled the curtains closed and shifted her. When

I returned from a moment in the kitchen she was in a cloaked position with wings half-open like a giant moth, clinging motionless to the division of the curtains high up while she peered through it, with one widespread foot clinging onto each side.

1 FEBRUARY

About 9:00 p.m. last night Mumble was fidgeting around in the living-room, clearly feeling disturbed, when she suddenly went bananas. I was standing close, trying to soothe her, when she hurled herself past me at the glass door leading to the balcony and started hooting furiously. As my eyes followed her I saw another tawny sitting on the balcony rail. Far from being put off by Mumble's abuse, he merely looked interested ("Well! *Aren't* you a spirited little thing . . ."—in my mind's eye I could see slicked-down hair, a pencil-line moustache, shined-up shoes, a bunch of flowers in one foot, and a box of chocolate-covered mice under his wing). He took off when he saw me, but not before hesitating for a second, which I found mildly insulting. Clearly I am not the only one around here who thinks Mumble is cute.

4 FEBRUARY

After several disturbed nights, I think I've found a way to stop Mumble screeching back at her hopeful suitors, whom she gives every sign of regarding with furious territorial resentment. Quite counter-intuitively, what seems to do the trick is simply letting her fly loose around the flat all night. Perhaps it's the feeling of being shut in her kitchen cage and unable to defend her patch that drives her mad when she hears another owl close by? I've taken to feeding her earlier in the evening, and keeping an eye on her mood as night draws on. If she seems quiet and unruffled, I shut her in the night cage last thing, but if she's restless and noisy I just leave her loose to roam, with the freedom of the living-room, kitchen, bathroom, and hallway. This seems to have curbed the excessive night-hooting, and when I get up in the morning I find her dozing on top of the bathroom door.

When I do shut her up for the night, I've taken to covering the cage with an old blanket when I go to bed, and this also keeps her much quieter. I feel obscurely guilty about this, but it's necessary for the sake of the neighbours' sleep, and thus for Mumble's security. In the morning, when I lift it back and open the door, she jumps first to her doorstep perch for a brief "good morning" nuzzle, and then right up onto my shoulder, where she stays for a few moments. While I move around putting the kettle on and making coffee, she takes her time to wake up and get over the dopey effect of being in the covered cage.

19 MARCH

I went into the balcony cage at about 8:00 p.m. to bring her in. She was already wide awake and seemed fluffed and agitated, and as I went onto the balcony I heard a hoot nearby. When I got into the cage I held the basket open for her in the normal way, and waited for her to make up her mind to jump in—when suddenly a damn great Tawny Owl flew slowly past the balcony in a controlled glide, as close as he could get, looking in at us. This was at an altitude of certainly ninety feet above the street. Mumble shrieked with fury, and EXPLODED up and down the cage, while I ducked and did my best to crouch out of her way—not easy, inside what is essentially a large wardrobe. It was a couple of minutes after lover-boy had disappeared before she calmed down a bit, and I could talk her into jumping into the basket, with much indignant shaking of feathers and muttering.

All the excitable dialogue over the past couple of months has at least given me plenty of opportunity to note down the variations in her vocabulary. These seem to include six basic types of call, though she may go off into various jazz riffs around some of them:

(1) The normal daily conversational treble "*kweeps*" and alto croons, usually on a rising note.

(2) The long, fluting "Indian war-whoops," upon hearing me arrive when she's in her cage, or as a cautious interrogation when she goes into a dark corner or hole: a flat, monotonously unvarying "*w-o-o-o . . . w-o-o-o . . . w-o-o-o . . .*" in a minor key.

(3) A chitter of annoyance or frustration, when I'm slow to feed her, or when I insist that she do something that she doesn't want to—like getting into the transit basket in the morning before she feels inclined to do so.

(4) A sort of "whistling kettle coming to the boil" as she pumps herself up in indignation, e.g., when she notices a pigeon lurking around. This starts with a crouch, a puff of the feathery throat, and an

interrogation that quickly turns dangerous: "skwer? . . . *skwer? . . . SKWER! . . . SKWERKK!*"

(5) The regular challenge call for other owls: "*HOOO! . . .* hoo, hoo-hoo *HOOOO!*"

(6) "Kee-*wikk*! Kee-*wikk*!"—the usual response to hearing another owl. My occasional impression that she is using it instead of the hoot for an initial challenge is almost certainly because she is actually replying to a stranger's call that is far beyond my earshot.

She played a riff on No. 6 at about 10:00 p.m. one night, when she was sitting on the windowsill. One moment she was quiet and serene, the next she was bobbing and peering at something outside, and making a sharp, repeated "*wuk! wwukk! WWUKKK!*" This did not develop into any other sort of crescendo, and after a while it died away into a quiet, bitching mutter on a slightly different note, very spiteful and bitten-off: "*wwakk . . . wwakk . . . wwakk . . .*"

☉☉

AMONG HUMANS, NO one was prepared to brave Mumble's moods more courageously than my office assistant, Jean. In early June 1979 I badly wanted to go away for a week, accompanying brother Dick on a D-Day thirty-fifth anniversary tour of Normandy by enthusiasts with about a hundred old military vehicles, ranging from motorbikes and jeeps to an M10 tank-destroyer. It was obviously out of the question to take an owl on a foreign holiday in the back of a vintage armoured scout car—even British eccentricity has its limits. Providentially, at just that same time Jean needed somewhere to stay for a few days, and she bravely volunteered to owl-sit for me.

I conscientiously explained the various challenges involved, but she was sure she could handle them. We had a twenty-four-hour handover period, and the two were introduced. A photo of Jean sitting nervously with the owl on her leg reminds me that Mumble was prepared to tolerate her presence without gratuitous violence, but both of them look pretty suspicious, and I remember feeling anxious about what might happen when they were alone together for days on end while I was somewhere in the *bocage* and far beyond reach of a phone call (remember, this was long before mobile phones or e-mails).

June 1979: Jean and Mumble size one another up during their twenty-four-hour familiarization period before my departure for Normandy. Neither looks particularly relaxed, but Jean's courageous week as an owl-sitter passed without excessive drama.

I showed Jean where the rations for humans and owls were kept, and watched while she gingerly practised extracting one of the pallid little corpses from the freezer, thawing it in a bowl of hot water, and later dunking it in a dietary-supplement powder. (This preparation, marketed as "SA 37," contained various minerals, vitamins, and trace elements, and was recommended for all types of pets at times of stress or vulnerable health. I had no reason to think that Mumble fell into either of those categories, but I had noted from the handbook that she should be starting her summer moult any time soon, and I wanted her to be in top condition for this presumably trying experience.) I carefully demonstrated the workings of the transit basket and the Double-Reciprocating Owl Valve, and left Jean equipped with protective headgear, clothing, goggles, and gloves. It must have taken a lot of nerve for her to venture into the balcony cage for the first time, but she was a very game girl.

Normandy was duly liberated. The first evening in a bar in Sainte-Mère-Église was memorable; I remember the *patron* getting Granny out of bed to help fry chips for fifty. I also have a vivid recollection of one of our number—suddenly overcome with tiredness and emotion—being passed horizontally towards the door above the heads of the packed crowd and laid down tenderly in the fresh night air, with a curbstone for a pillow. Equally unforgettable, in a different way, was a sunny morning spent mooching around on Utah Beach in search of souvenirs, when to my astonishment I found not only cartridges, but even the collar of a GI's field jacket still trapped between two small rocks in a tidal pool.

Thereafter, however, it rained, and kept on raining. The only consolations were that at least I had discovered Calvados, and that the charming girlfriends of two friendly Dutchmen following us in an old Dodge ambulance were much better-looking than my male companions. The underpowered White M3A1 Scout Car—christened "Pig Pen," from its squalid state while accommodating six permanently dank and grubby Englishmen under field conditions—broke down enough times to keep Dick happy (for him, it was the challenge that counted). Whenever this happened I found myself—weirdly like my owl—instinctively drawn to the highest perch I could find, where I sat smoking glumly while I watched my brother's legs or backside sticking out from under various parts of a sullenly unresponsive four-ton assembly of olive-drab steel.

How he ever got us onto the ferry home I will never know. The sun came out again on that final morning, but all of us except Dick were savagely hung-over, and Pig Pen became stranded within sight of the Boulogne docks. I remember sitting on the curb with one of my crapulous fellow Pigs, sharing a carton of chips bought by pooling our very last francs, while we waited for Dick to come up with another miracle—which, of course, he did. As I recall, this final triumph involved trading the loan of a spare wheel against the promise of a surreptitious cross-Channel passage for a wartime flamethrower that somebody had acquired as a souvenir.

We struck lucky. When we docked at Dover the customs officers seemed to be picking vehicles to search on the eeny-meeny-miney-mo principle, and the DUKW ahead of us had to disgorge about a dozen protesting passengers down a ladder. After taking one look into the hull of Pig Pen at the pile of malodorous bodies sleeping it off on a floor of muddy steel, crumpled ponchos, and food debris, the customs guys wearily waved us through, all innocent of what lay beneath.

When I finally got home Jean appeared to be unscarred; she reported no episodes of bloody mayhem, and Mumble looked in fine fettle ("New, improved Mumble—now with SA 37!"). It turned out that the biggest problem had been a bachelor neighbour of mine. After meeting the slim, blonde Jean in the lift one day he was insistent on pursuing the acquaintance, and she had had to spend some time discouraging him. Her side of these conversations was necessarily shouted through the closed front door while she struggled to think of a convincing explanation for her reluctance to open it.

◎◎

DESPITE HER BEHAVIOUR in the mating season, I remained uncertain about Mumble's possible reactions to other owls in the longer term, and I thought that the easiest way to check this out without taking risks was to drive her down to Water Farm for a weekend and introduce her to the docile Wol. So one Saturday morning I packed my kit, and put the transit basket open on the kitchen table. Mumble had been fairly listless that morning; now she took one look, then jumped into the basket of her own accord. Once I got down to the basement garage and inside the car I put the basket down in the plastic-lined rear cargo area and opened it. I noted that she emerged with a slit-eyed, stoned expression, and made her way up to my shoulder with a kind of blind determination. (By day, tawnies are hard-wired by Nature to sit beside things, so my shoulder and head were doing the job of the base of a roosting-branch and the tree trunk next to it.)

And there she stayed, for at least forty-five minutes—all

Self-portrait in the bedroom mirror; Mumble has
just spotted something more interesting outside the
window. When she saw herself in a mirror she
never reacted as if she was being confronted
by another owl.

the way out through the Saturday shopping streets of South
London, with their constant traffic-light halts and stop-start
crawling along between crowded sidewalks. We drove slowly
for miles before we reached the divided highway that took
us out to the Kent countryside, yet as far as I could tell, *not a
single person noticed that I had an owl on my shoulder*—not

even passengers in cars stopped beside me, whom I would at least have expected to do a brief double-take before deciding their eyes were playing tricks. She nibbled my ear a couple of times; crapped once down the plastic sheet over the seat behind my shoulder; then decided to go back into her basket, where she stayed for the rest of the two-hour drive. She didn't make a single sound the whole way.

Mumble stayed dopey, malleable, and a bit off her feed for the whole weekend she spent in one of Dick's spare aviaries. She seemed riveted by the sunlight on the grass and the duck pond, and by the chickens and ducks nearby. When Avril brought Wol out on her fist, and showed them to each other through the wire mesh, Mumble adopted a hunched, cloaked posture, just as she did when she heard a wild owl outside the flats. But she didn't get hysterical, and she didn't challenge him; could she have recognized that she was on his turf? Anyway, the experiment was inconclusive.

I would repeat it more extensively the following Christmas, when we went down to Water Farm for a five-day family holiday. The spare aviary was in an exposed position and the weather was dreadful, so I was grateful for my nephew Graham's help in trying to fix a wildly flapping plastic sheet to shelter the roof and one side of Mumble's quarters against the rain. (I had long ago learned that she was quite unworried by bad weather. If caught in her balcony cage by a thunderstorm she seemed to positively relish it; she would emerge from her hutch to crouch on her furthest front perch in the blowing drizzle, admiring the lightning-flashes like a child at a fireworks display.)

The adjoining aviary now housed a semi-wild pair of Tawny Owl siblings, so the Christmas visit promised to be interesting. In the event, Mumble seemed perfectly happy so long as she could see these neighbours, and she spent some time chatting to them with interest but no apparent hostility. They seemed the more anxious; they often hid from her, and when Mumble heard them moving around out of her sight she got suspicious and watchful. She ate her chicks with a hearty appetite, however, and took a bath despite the lousy weather. On several nights the three owls sang together, Mumble making the most noise. Since they were only a couple of feet apart and usually in mutual view, it did not seem likely that they were challenging each other, and more plausible that they were collectively sending warnings to more distant owls.

After we got home there were slight but unmistakable signs that Mumble's routine had been thrown by meeting these temporary neighbours. She was never rough, but for three or four days she was a bit stand-offish, and she demanded food, and hooted, at the wrong times and places. I consoled myself for my slight and fatuous feeling of neglect by reflecting that this seemed to confirm that if anything happened to me then she could probably be introduced to another aviary without much distress.

◎◎

DESPITE HER ALLEGEDLY polite behaviour towards Jean during the owl-sitting episode in June, as the summer and autumn of 1979 had passed it had become clear that Mumble was getting ever more territorial. Her behaviour towards me didn't change, and it was still sometimes possible to bring her in to meet an insistent visitor after they had already

been installed on the living-room sofa. However, if somebody arrived when she was already loose in the flat she had
started to regard it as an intrusion on ground she had occupied, and she might fly for their scalp. Tin hats were all
very well, but on a couple of occasions I had to catch her
quickly and shut her inside the kitchen. There she would
leap around and scratch at the other side of the glass door,
with furious glares and hoots. This was embarrassing; some
visitors chose to take it personally, and it also greatly complicated the process of making coffee for them.

It was during the autumn of 1979 that Mumble's behaviour towards guests became terminally intolerant, and intolerable. One evening I was cooking supper for Graham,
who was passing between the kitchen and the living-room
table with cutlery, wine, and so forth. Mumble was sitting
on her door top, watching him pass below her. In all previous encounters with him she had been friendly enough, or
at least politely distant, but on this occasion he didn't like
the way she was looking at him: "There was a definite sense
of radar lock-on." The next time he walked past he felt a
sharp clout at the base of his skull. Startled, he put a hand
up, and it came away bloody. Seeing Mumble back on her
door, measuring the range once again, Graham felt mixed
emotions (very fast). On the one hand, she was fluffy and
cute and was his uncle's treasured pet; on the other, she
had just drawn blood and was obviously about to try again.
He just had time to snatch up an empty cardboard box to
use as a shield; it worked, but Mumble immediately circled
for another run.

When his yell brought me sauntering in, still with a

Mumble at about nine months old. The evidence of
my consuming coffee by the pint, and the fact that
she is free in the living-room, suggests that this was
taken on a weekend morning, when we did
much of our daytime socializing.

wooden spoon in one hand, my reaction on seeing him
holding a cardboard box above his head was apparently less
than helpful. He recalls a lot of tutting, and "Well, she's
never done *that* before"; I am ashamed to recognize the infuriating manner of fond dog-owners, whose tone implies
that victims of their slavering mutts have only themselves to
blame. Mumble was duly scooped up and re-inserted in the

balcony cage; but Graham remembers that his relief was definitely tinged with sadness, that he could never again share the same room space with this beautiful wild creature.

Despite this clear evidence that my owl was now—perhaps rather belatedly—fully in the grip of the adult instinct to defend her hunting and nesting territory, I confess that I remained briefly in denial. That came to an abrupt end soon afterwards, on a day when my friend Bella visited. The former Mumble had been happy to accept at her hands a certain amount of "coochie-coochie-coo," and this time Bella reached up to her on the door top as usual. Mumble immediately dropped on her head like a feathery brick, all talons extended. As I checked her scalp for cuts, Bella let me know—with the directness proper to her birthright in the Northern Caucasus—how she felt about this treachery on the part of something that she had previously regarded as an animated fluffy toy.

The lesson was clear, and final. Mumble was no longer an unpredictable but generally adorable pet to be shown off and shared, but a grown, territorial, dangerous, and strictly one-man bird. From that day until the end of her life I could never allow anyone but myself in the same room with her. Many years later Dick and I performed an experiment (with protective headgear for him). My brother and I are superficially alike—similar in height and general build, both bearded—and as an experienced falconer he projects no nervousness when he is around birds of prey. We spent time together in the same room as her uncovered night cage so that she could get used to him. As soon as I opened the cage she attacked Dick, but then allowed me to handle her to put her away again. Whatever bond she and I had, it was between us alone.

∞

I OCCASIONALLY RELENTED over my rule about never letting Mumble into my bedroom or office. The bedroom was the only place where I had a large enough wall mirror in good enough light to attempt to photograph her sitting on my shoulder. Since the room had very little space for anything other than the double bed and a chair, and offered neither attractive perches nor a window view any different from that in the living-room, she normally showed little interest in it.

The exception was the first occasion when I happened to be changing a duvet cover while she was free in the flat. That always awkward exercise was doubly so in this very confined space, and was physically impossible without leaving the bedroom door open. When Mumble caught sight of me struggling with the billowing cotton she naturally interpreted it as a new game invented entirely for her amusement. Given her fetish for holes and tunnels, it was inevitable that the large, contorted bag of the half-replaced duvet cover would attract her at once. At the first opportunity she swooped down and scurried inside it; then, whooping like a Comanche, she tried to force her way down to the furthest corner, between the duvet's interestingly squidgy surface underfoot and the thin, translucent tent resting lightly on her head and back. It was some time before I could extract her and banish her to the kitchen, and not before her claws had left their mark on the bedclothes.

Like any fond parent, I found it hard to apply the rules consistently, and in a weak moment one day when I was working at home I let her come into the office with me. (There was always a temptation to expose her to some new experience, simply for the interest of watching how she reacted.) I can't recall that first occasion exactly, but I suppose she was being particularly delightful one morning, and I simply thought, *Oh, what the hell—what harm can it do?* Of course, once I had first weakened I did so again; I soon lost all moral authority, and Mumble found the office a much more interesting space than the bedroom.

It was large but dimly lit, with my desk under the window at the end looking out onto the balcony and her cage. It had another large built-in wardrobe with a sliding door, just like the one she enjoyed so much in the hallway, though this one was mostly full of old military uniforms. There were stacks of bookshelves along the other walls and in a free-standing pilaster, and in the corner was a life-sized dummy wearing a Foreign Legion parade kit. (To my relief, Mumble apparently found the large green-and-scarlet epaulettes on its shoulders uncomfortable as perches—these collector's items, the gift of a veteran, had survived two wars, and I didn't want them splashed with owl-crap.) The wardrobe and bookshelves offered many intriguing crannies where she could creep between and behind things, and when she came into the office she generally didn't flit around for long before finding herself a nice dim den to settle in. She was no trouble while I was working at editing typescripts with a pen; the problems started when I began typing.

To the digital generations who are too young ever to have

seen one in action, I should perhaps explain that a manual typewriter with metal keys made a much louder clacking noise than the plastic keyboard of a computer. Moreover—and crucially, in this context—a sheet of paper was tucked around a cylindrical carriage mounted across the top of the machine, which moved steadily from right to left as the keys were struck. When you got to the end of each line of type the carriage stopped with a slight chiming noise, and you automatically reached up with your left hand to a lever on the end of the carriage and slammed it hard back to the right again, which rotated it to the next line space. To summarize, this device had a sheet of paper waving out of it, made a rhythmic noise, and was in constant movement from side to side, punctuated by chimes and exciting rushes ending in a crashing sound. What more could an adventurous young owl possibly desire?

The first time Mumble decided to investigate she came from behind me, simply hurling herself into the machine talons-first and with wings upraised, as if she were jumping into a potted plant (a favourite game of hers). I'm a fairly fast typist, and when I am writing I concentrate hard, so when she arrived at speed in the central well of the machine she got a couple of key-taps under the tail before I could react. She found the rise and fall of the long keys under her feet intolerable, especially at a moment when she was trying to concentrate on biting the paper in front of her face, so things got a bit flappy and indignant before she was back on a bookshelf, thoroughly disgruntled.

Anybody would think that this would have put her off, but she was nothing if not obsessive in her interests. The

A classic "cottage loaf" pose, on the telephone table
in the hallway. For some reason she always regarded
the phone with some suspicion and objected
vigorously whenever I used it.

sideways progress of the attractively waving sheet of paper
proved irresistible, so she simply had to figure out an
approach that kept her free of the annoying thrashing of
the keys. This did not take much ingenuity, and before long
she was landing from in front of me, directly onto the end
of the carriage beside the paper. The first few times she did
this I stopped typing and shooed her, chittering, off the

machine. That only made her more determined, and she stubbornly returned again and again, until I lost patience and ejected her from the office.

We repeated this battle of wills on numerous occasions, and she made progress. I would keep right on typing, and after she got used to it she found that she enjoyed riding the carriage along its right-to-left track. This was apparently exciting enough, and she usually gave up her attempts to savage the paper. Naturally, every time I slammed the carriage back to the right she jumped into the air, but she soon learned to simply hover for half a second until its crashing arrival at the end of the track, and then descended onto it again for the next ride. I won't pretend that I got much copy written while she was playing this game, and after a while I had to either distract her or banish her. Any sane person would simply have stopped allowing her into the office, ever, but I must confess that I found the spectacle of her riding the carriage rather endearing, and I never could bring myself to impose a permanent ban.

MUMBLE CONTINUED TO take an intelligent interest in the written word, and when I was sitting reading with a newspaper on my lap she might suddenly arrive out of nowhere, landing in the middle of it with a crash and happily kicking holes in it. When I was lying on the sofa she would sometimes land unexpectedly on my chest and walk up to my face, to investigate my beard. One summer evening I was stretched at my ease with a book propped on my chest; Mumble was off about her own concerns somewhere, and I was completely absorbed in my reading. Suddenly, and absolutely without warning, she landed heavily in the narrow space between book and face. My protest left my brain as "Good *grief,* Mumble!" but reached my ears as "F'noog *f'neef,* Unguh!" since her fluffy front was pressing hard against my nose and mouth. She apparently construed the resultant burst of warm air up her petticoats as a physical liberty, because she bent forwards and carefully bit me on the bridge of my nose.

In the autumn of 1979, when she was about eighteen

A glass of wine, a cheroot, music, and a contented
owl—what more could one wish for on a quiet
evening at home?

months old, I noted an unwelcome change in Mumble's habits. Sometimes, instead of (or after) sitting on my shoulder, she would take up position on top of my head. I suppose that the attraction was extra height and superior all-round visibility, which was fair enough; but it involved the frequent shuffling around of sharp claws to adjust her balance, and the subsequent take-off kicks could be quite painful.

Specific occasions when she chose to do this included any time when I used the telephone in the hallway. This seemed to provoke in her a positively childish competition for my attention. She might be sitting at a window gazing serenely over the roofscape, or dozing contentedly on top of a door, but if the phone rang, or I dialled out, then within a moment she would arrive on my scalp. She would squeak peevishly, pecking downwards at the handset or my ear, and then jump to the crook of my arm and try to bite through the dangling spiral cable. Callers who were ignorant of my domestic arrangements sometimes found the resulting three-way conversation confusing. I hesitated to share with them the fact that I was conducting it with an owl sitting on my head, for fear that the more conventional-minded clients might think such behaviour unprofessional.

◎◎

I HAD ALWAYS left Mumble a shallow water dish in her balcony cage, knowing from Avril's experience with Wol that owls enjoy the occasional bath. She seemed to do this at least once a week, though it was hard to be sure. The first time I saw her doing it she reminded me of a human getting into a bath of uncertain heat. She stood for a few seconds on the rim of her dish, then stepped down into it with dainty caution, one foot at a time. She thought about things for a moment, standing with the water half-way up her lower legs; then she slowly settled downwards and forwards into it, until she was lying on her front with the water now half-way up her sides. She fluffed her feathers out a bit, wriggled, and did a few gentle push-ups, waggling her tightly folded wings slightly. She kept these furled up, but then began clapping them against her sides more vigorously while ducking her face down into the water and up again, splashing droplets up and over her back. After several of these wing-waggling sessions she lay still for a while, obviously enjoying the "soaking" sensation. Then she clambered to her feet and stepped carefully out, to begin the lengthy process of shaking herself dry and grooming her feathers.

This interest in water extended to times when she was free in the flat. When I was washing dishes in the kitchen sink she would sometimes fly to my shoulder, looking down with apparent fascination at my hands splashing among the suds and plates in the washing-up bowl. She was clearly trying to decide whether or not to jump down there and join in the action, but she never quite nerved herself to do it. However, on one really adventurous bath night it became apparent that she had been keeping her eyes open until I left water in the otherwise empty bowl.

I was relaxing in the living-room that evening, not thinking about what she was getting up to, when I heard a *flopp!* like a soaking-wet dishrag falling on the lino floor of the kitchen. As I looked across to the kitchen door, something appalling came waddling slowly into view around the corner at floor level. Mumble had clearly been right under the water, and for some time, because her head was as completely soaked as the rest of her body. A longer beak than I had ever seen, and a pair of madly staring eyeballs, were sticking far out to the front of a tiny head covered with a thin Goth hair-style of long black spikes ("*Baby!* I hardly *recognized* you!"). Her body was a dark, bedraggled mass of ratty tails, like those handfuls of soiled sheepskin that you find caught on barbed-wire fences in wintertime, and her wings looked like a wrecked umbrella in a storm-drain.

Muttering and complaining under her breath, she came hopping and clicking towards me across the floor, making completely pointless wing-flaps every few steps. She was so heavy that she could not even jump up to my wrist. She had to find a ladder of graduated footholds so that she could climb—from floor to footstool, from there to my knee, and then laboriously up my chest to my shoulder, still bitching and moaning all the time. Once there, she tried to shake half a pint of water out of her wings, but her balance was so bad that she nearly fell off again, and she made me jump by locking her talons painfully in my shoulder.

I got carefully to my feet, and she clung on tightly while I walked her slowly back into the kitchen. The ceiling strip-light was only a foot or so from the top of a stack of free-standing shelves. She climbed slowly and unsteadily up the inclined bridge of my arm until she could reach the top shelf. Again she tried to shake, but the flopping weight threatened to carry her over the edge. So there she stayed, as close to the warmth of the light as she could get, drying out by slow degrees for the rest of the evening.

This was obviously going to take hours, so I left her to it.

About four hours after her total-immersion experience in the
washing-up bowl, Mumble still looks fairly ratty.

From my chair in the living-room I couldn't see her, but a
score of times I heard the rattle of her shaking her sodden
feathers in furious, ten-second bursts. If I craned backwards
I could see her shadow on the kitchen wall, huge and night-
marish, as she turned herself inside out and upside down in
a frenzy of flapping and grooming. When I went into the
kitchen after a couple of hours to see how she was getting
on, she was leaning slightly forwards into the light with her

thin racks of separated black flight feathers half open; she was an evil sight, and she seemed to know it. Her chances of achieving flight were about the same as one of those Edwardian cartoon "intrepid birdmen" with strapped-on wings, about to fling himself off a clifftop with deranged confidence. ("It'll never work, Mumble—stay by the sunlamp.")

That night I served the chicks in her open cage and left the kitchen light on, so that she could make up her own mind where she wanted to spend the hours until sunrise. She looked in pretty good shape by the next morning; the experience didn't put her off occasional light bathing in the washing-up bowl, but she never again tried swimming underwater.

ANOTHER HABIT THAT came as something of a surprise to me was her enjoyment of sunbathing. I was vaguely aware, of course, that on sunny days most birds occasionally like to get belly-down on the ground in a patch of dust and flap their feathers about. The dust helps clear out parasites, and the sunlight is a necessary source of vitamin D. But for some reason I had never connected this practice with night birds like Mumble, even though I had quickly learned that she was quite happy in the sunshine.

One summer weekend I was reading in the living-room with the balcony door open when I heard a loud "*splat!*" from the cage—the sound of Mumble jumping down to the newspaper carpet. After a few moments I got up and took a peek around the corner onto the balcony, to satisfy a vague curiosity about what she was up to. For a micro-second my heart lurched: Mumble was lying flat on her belly on the

Mumble sunbathing on the floor of the balcony cage. She
is lying flat on her front, with her face turned up to the
sunshine, and her wings, tail, and body feathers
spread out as widely as possible.

floor. But almost immediately I sensed, from the relaxed nature of the slight movements that she was making, that she
had taken up this pose deliberately. She was lying flat on her
front in the largest patch of sunlight that she could find, with
her wings spread wide, her neck bent back, and her face
pointing directly upwards into the sunshine with closed eyes.

Before I turned quietly away and left her to enjoy it, I
noticed that—oddly—the sunbathing session seemed to involve much the same sort of facial transformation as when
she snapped onto the alert when spotting a pigeon taking
liberties outside the window. Why this should be, when the
one activity was presumably a sensual pleasure and the other
a prelude to combat, seemed baffling, but there was no mis-

taking the resemblance. The skin of her head looked tight, and the close packing of the feathers gave her a "pinhead" look. The furry wedge of feathers between her slitted eyes protruded and spread wider sideways, covering the inner top part of the eyes so that they appeared tilted and further apart. This squared-off or Chinese expression was absolutely distinct from her usual appearance in repose.

◉◉

FOR THE FIRST time, in November 1980—so when she was about two and a half years old, and at completely the wrong time of year—I noticed "broody" behaviour that finally convinced me that Mumble must indeed be a hen bird. I found her sitting in the semi-darkness on the hallway table on a large dish-shaped ashtray. During our first months together I had often found her lying flat on her front for a rest, but this was quite different. She had arranged herself exactly like a chicken sitting on its eggs—lying on her front, head back and tail cocked up, with her body feathers puffed out sideways and her wings folded protectively around the "nest." She seemed dozy, and made sleepy cheeping noises when I stroked her. Very occasionally over the hours that followed she would stand up, shuffle sideways around the rim of the ashtray while pecking vaguely at it, and then fluff herself up and settle down on it once again with a shake of her skirts. She kept this up, off and on, for about eight days. (Although I kept an eye out for this behaviour the following spring, and in the years thereafter, I very seldom saw her repeat it, and never for days at a time like this first occasion.)

DIARY: 30 DECEMBER 1980

Since getting back from the Christmas holiday at Dick's, and remembering both Mumble's broody behaviour last month and her previous response to other owls, I've been thinking about what she has missed through being born into captivity. There is obviously something about caging a bird that we instinctively see as wrong, because the very idea of flight represents freedom to us, so I had better keep myself honest by thinking this through every now and then.

Birds evolved to fly because flying offered them an advantageous edge in the great competition of life. The price for this is that flying demands a considerable investment both in the energy to power strong muscles fuelled by a fast-beating heart, and in the self-maintenance of a complex, sometimes fragile body. Science teaches us that some originally flighted species of birds have subsequently become flightless, because their environment changed in ways that removed the need for this specialized behaviour. Such species "chose" to give up flying when it ceased to deliver real advantages, and so they gradually lost the physical ability. Logically, they would not have done this if the "freedom of the skies" had been necessary to their mental health.

Obviously, the act of flying plays the central part in the ability of most bird species to sustain life, and nobody who has watched the dazzling aerobatics of, say, a peregrine falcon or a plover can believe that it does not take some kind of satisfaction from exercising its strength and agility. But not even all raptors depend upon flying to the same extent, and woodland owls like the tawny are among those that Nature has designed for relatively brief, low-altitude flights. In terms of human military aviation, we might think of them as "short-range vertical/short take-off and landing ground-attack fighters" rather than as "air-superiority interceptors" like peregrines. Mumble is a Hawker Harrier, not an F-16.

If she had been born in the wild (and, of course, if she had been among the small minority of the 1978 fledgelings to survive their first couple of winters), then Mumble would spend the great majority of every twenty-four hours roosting in one or other of her preferred trees. During the daylight hours she would sleep or doze, tucked in concealment against the tree trunk, while she digested last night's kills. Before nightfall she would rouse herself, bring up the resulting pellet of waste matter, and preen her feathers in order to maintain her flying and protective surfaces in good condition. She might begin the night's working shift by patrolling the boundaries of her territory, issuing a few challenges if she suspected there were interlopers nearby. She would then move from tree to tree by short flights, sitting for alternating spells on these regular perches while she listened and watched for the movement of prey animals.

As soon as she had killed a sufficient "bag" for the night—which she might achieve relatively quickly, or not until some hours had passed, depending upon the year and the season—she would return to a roosting tree. There she would eat and begin to digest her meal. At about dawn she would slip once again into a dozing state, remaining motionless among the leaves so as to stay hidden from daytime birds. Except during the springtime months when she was helping to raise a brood, she would never expend any more energy than she needed to catch enough food to sustain her own life.

It is natural that Mumble chooses to sleep or doze for much of the twenty-four-hour cycle—this is innate behaviour, not a response to captivity. She couldn't show any less evidence of "cabin fever" if she tried, and her attitude to life seems to be more or less that of a lazy cat. Since she is probably fed better (and certainly more consistently) than she could feed herself in the wild, why shouldn't she be lazy? And since she seems calm and friendly most of the time when she is awake, I can rationalize any residual "jailer's guilt" by telling myself that she is, for the most part, living a contented life. Freed from the nightly dangers of the woods and roadsides, this life will probably be a great deal longer than it would have been if she had been hatched in the wild.

The guilt cannot be entirely washed away, of course; I have a human sensibility, not an owl's. I cannot forget that captivity denies her, if nothing else, the fundamental experience of breeding, but I find that I can live with this. Above all, nobody kidnapped her from a wild life, so she has no concept of greater freedom; and if she is often caged, then at least she has never been tethered.

6

◎◎

The Driver's Manual

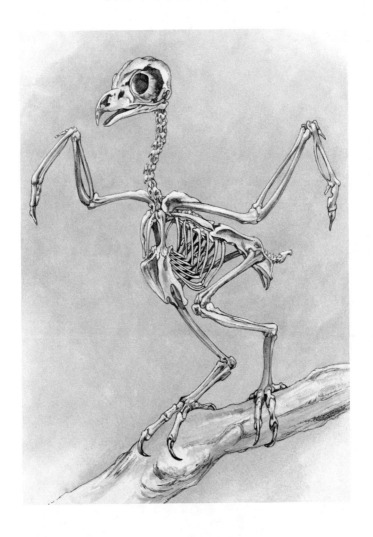

WHILE IT WAS Mumble's looks and behaviour that captivated me, acquiring an elementary knowledge of the "owner's handbook" still seemed in order. I never became even an amateur ornithologist, but I still felt the inborn male need to understand what I was looking at. Even though my grasp of the scientific principles of the internal combustion engine is shaky, I still want to know roughly what's under the hood, and how the major moving parts interact. Consequently, this chapter may be regarded as a layman's brief inspection tour around Mumble, with the manual in one hand. What I discovered while doing even some basic reading was impressive.

☯☯

THE TAWNY OWL is rather charmingly described as "portly" in the 1943 *Handbook of British Birds*, and the first surprise was the enormous difference between Mumble's external appearance and the "cutaway diagram" (see the drawing on page 167). She was a miracle of concealed compression; although her normal comfortable squat made her look like a relaxed bundle of feathers about ten inches tall not counting the tail, her skeleton, if stretched out from head to talons, would have been at least 50 per cent longer. The most striking revelation was the length of her snaky, S-shaped neck, which was made up of about twice as many cervical vertebrae as a human's. When she was at rest her neck was completely invisible, hidden by a sort of telescopic effect of the deceptively thick, loose ruff of plumage between the "ball" of her head and the bulk of her body. In repose her neck was arranged like a swan's, and its concealed length obviously explained her ability to twist her head around to such extreme angles. It also contributed to a trick that she demonstrated far less often, but which I found so amusing that I was sometimes tempted to tease her so that she would keep doing it.

It is only in recent decades that human engineers have come up with a really efficient gyroscopic stabilizer for incorporation in tank turrets (bear with me—there is a point to this). If you watch film of a modern battle tank rolling fast across rough ground during a firing exercise, you will notice the uncanny way in which the gun remains aimed steadily at a fixed point even though the tank's hull below it is rolling and pitching in all directions. Achieving this took half a century and cost many millions, so imagine my delight when I discovered that Mumble could do it automatically.

One day I happened to lift her off her tray-perch while she was concentrating on something she had spotted outside the window. As she stepped backwards onto my hand and I lowered my arm, her head remained immobile—I don't mean in relation to her body, but *at exactly the same height and angle in mid-air* as when she had been up on the perch. As my hand and her body descended, her neck simply deployed upwards out of her shoulders, the feather ruff stretching and growing narrower, so that her head remained fixed in all three dimensions while her body dropped several inches below it. This looked so weird that I instinctively raised my hand again—her head simply stayed where it was while her neck steadily disappeared, telescoping smoothly down into her shoulders, with the feathery ruff fattening out around it again.

I need hardly add that I immediately fell prey to a mischievous temptation to play "owl yo-yo"—lowering and raising my hand, and watching her head remain fixed in mid-air while her neck extended and contracted below it. I did this several times, giggling foolishly, until Mumble got fed up with this childish game and took off.

◎◎

WITH A LITTLE imagination, it is possible to figure out from a drawing how a bird's skeleton evolved from that of a reptile all those eons ago in the Upper Jurassic. After some 160 million years it still recognizably has a head, a neck, a torso, a pelvis, a tail, and four limbs, all more or less where we would expect them to be. However, a simple outline drawing of the skeleton does not tell us the half of it.

When I got stuck into the textbooks, I learned that Mumble's body was designed—logically enough—with the primary object of achieving a high power-to-weight ratio. Having once been an aviation journalist, I couldn't help seeing her as Nature's equivalent of an aircraft that has much of its structure built from drilled-out aluminium and moulded carbon fibre, to reconcile strength with light weight. To provide the thrust to get it into the air and flying at operational speed, this structure was powered by a high-revving "engine" (her heart and lungs), consuming large quantities of "fuel" (oxygenated blood) circulating at high speed.

When I picked her up she seemed too light for her

apparent bulk, and this was not only because most of the "visible Mumble" was actually made of feathers and the air trapped between them. Her lightness was also due to the fact that many of her bones—parts of her skull, vertebrae, breastbone, humerus or "main wing spars," her ribs, pelvis, and legs—were actually partly hollow, with air spaces inside. You would think that this made them dangerously fragile, but they had an internal bracing of little bone struts across the hollow shafts. The arrangement of her innards also showed some emphasis on saving weight: the designer had left out some of the squishy bits and liquid that we have, and in some cases of paired organs only one of them was fully developed and functional.

Many of the bones of her body—her massive breastbone, the shoulder blades, lumbar vertebrae, ribs, and pelvic girdle—were fused together at their extremities, so as to create a rigid box structure protecting her internal organs. The most prominent part of this was the big keel of her sternum, or breastbone, to which the powerful flight muscles were anchored; this was braced to her shoulders by two strong bones, to prevent the pull of the muscles collapsing the box of her torso.

Inside this protective "fuselage" structure, the power plant that enabled Mumble to generate the extraordinary muscular activity necessary for flying was a heart that was—relative to our body sizes—much larger than mine, and beating much faster. At rest, this amazing engine beat about three hundred times a minute (four times the human rate), and it pumped—again, relatively speaking—about seven times as much blood, at much higher pressure.

The "carburettor" that mixed the fuel from blood and

oxygen was a pair of lungs that were relatively smaller than mine, but that were linked to an extensive secondary air-circulation system; this got large quantities of usable oxygen into her bloodstream quickly, while also contributing to the buoyancy of her body. Birds have a network of (usually nine) internal air sacs—simplified extensions or "annexes" to the lungs. These act rather like bellows, moving fresh, oxygen-rich air through the lungs quickly without its getting mixed up with the stale "exhaust" air already inside the system. Eight of a Tawny Owl's air sacs are arranged in pairs down the sides of the breast and belly, and the ninth, shaped roughly like an upside-down triangle, is mounted centrally at the top. This last sac has pipe-like pouches at the top corners that extend out along the inside of the humerus bones; when Mumble breathed in, air passed through her lungs and air sacs and right up inside her wing structure.

Unlike mammals, birds have no diaphragm muscle, and they breathe by expanding their rib cage to draw in air. (This is why it is vital, when holding a bird, not to constrict its torso—if the ribs can't move, the bird can't breathe.) It is almost impossible to see an owl breathing, unless you catch a glimpse of the slight, regular movement of the lower back feathers between the folded wings.

◎◎

WHILE I WAS reading up on owl anatomy, I often found myself glancing back and forth between a book in my hand and Mumble sitting modestly beside me, and muttering things like "Damn me, sweetheart! There's a lot more to you than meets the eye, isn't there?" One of the first things to fascinate me was an explanation of the inner workings of her own eyes. Behind her innocent gaze and friendly blink, there was a great deal going on that I had not previously understood.

Owls' eyes have much the same basic components as our own. Behind the cornea (the transparent outer cover), the adjustable muscular ring of the iris (the coloured bit of our eyes) regulates the passage of light through the pupil or central hole (the black bit). In bright light the pupil shrinks, in low light it opens out, and this regulates the amount of light passing through the pupil onto a lens behind it. The lens focuses the light-images of what we see onto the retina, the "screen" at the back of the eyeball; the retina is connected by a bundle of optical nerves to the visual cortex of the brain, which interprets the images.

Despite the basic similarities, the arrangement of these components in an owl's eye differs significantly from our own. The most obvious difference is that its eye "ball" is not spherical at all, but a sort of truncated cone, supported by a ring of small bone plates forming a short, tapered tube. We might perhaps compare the shape of the eye roughly with that of a lightbulb, or—less prosaically—with an Apollo manned space capsule. The lightbulb/capsule's narrower, tubular end represents the cornea, iris, and pupil, with a very large, thick lens at its base. Behind the lens the owl's eye swells out in a blunt conical shape inside the skull, and at the wide end the broad, curved retina at the back represents the Apollo's heat shield. The front-to-back depth of Mumble's eyes was actually greater than that of my own, despite my much larger head. Because of its elongated shape, the eye cannot swivel around in its socket—there is simply no room inside the skull to accommodate any movement. Mumble's eyes remained fixed forwards and pointing very slightly apart, but her long, flexible neck compensated for this by allowing a huge range of head movement.

There are two types of photoreceptor cells in the retina at the back of the eye. "Rod" cells are sensitive to light intensity, and "cone" cells are associated with both colour vision and resolving power—the ability to distinguish fine details. The photoreceptors of daytime birds are about 80 per cent cones, which pick up colours all the way across the spectrum from red through yellow, green and blue to ultra-violet. By contrast, the owl's large retina, backed up by a reflective layer, is particularly rich in closely packed rod cells, which allow the eye to function well over a very wide range of light levels. The rod cells of Mumble's retina were packed more than five times as densely as those in my eyes, and the image projected onto her retina was 2.7 times as bright as that created on mine. Some experts believe that Tawny Owls may actually have the best low-light perception of any vertebrate creature, and that in this regard a tawny's eyes have evolved to the limit of theoretical perfection.

The cost of this high sensitivity to light is a reduction in "acuity" or resolving power, which is only about 20 per cent as good as that of some daytime raptors. At light levels at the very bottom end of their visual range, owls can make out only the coarsest contrasts, though such darkness occurs only under the woodland canopy; if they can see the sky, this always provides enough light for them to see things with reasonable clarity. Compared with our own eyes, those of owls have far fewer cone cells relative to rods, and their exact ability to distinguish colours has been a matter of debate. However, it has been established that Tawny Owls' eyes are, at the very least, sensitive to the yellow, green, and, to a lesser extent, blue wavelengths.

An owl's superlative ability to adjust its eyes to compensate for a wide range of light levels also works the other way around. The idea that they are blind in strong sunlight is entirely mistaken; they can see better by day than by night. When Wellington, my Little Owl, was in the dimmest light, his irises were narrow yellow rings surrounding big black discs; when he was out and about by day his pupils shrank to black dots in the middle of big yellow discs, but he showed no inclination at all to stay out of the sunlight.

Neither did Mumble—far from it—but because the whole outer surface of her eyes was very dark, their adjustment to light levels was virtually invisible. They were big, slightly protruding, and appeared to be a glassy black. It was usually impossible to tell the central pupil from the surrounding iris, but very occasionally, if sunlight struck her eye at a particular oblique angle, I could distinguish the faintest of colour differences between the black-brown outer ring and the jet-black centre.

The rims of her eyelids were coral-pink, and the eyelids themselves—top and bottom—were covered with what looked like the palest grey-buff fur. Both lids closed, but the lower lids seemed to close upwards further than the upper lids dropped. She closed her eyes when she was sleeping, or wanted to protect them—noticeably, when she was either eating or grooming herself. (When she was young I formed the impression that blinking was also a sort of greeting, so I used to blink slowly back at her.) When she wanted to wipe her eyes, or to protect them without closing

them, she instantly closed her "second eyelids"—the nictitating membranes, which slid diagonally across from the top
inner to the bottom outer corners (owls also close these
when just about to strike their prey). In owls, unlike some
other birds, these membranes are not transparent, and in
tawnies they have a milky-blue appearance except for a narrow transparent edge. If I happened to look at Mumble
while she was flicking them across, I got the startling impression that her eyes had changed colour instantaneously
from shiny black to a blind-looking and rather sinister
translucent grey-blue.

MUMBLE'S FACE WAS covered with very fine, soft feathers
that gave it a furry look. They were pale greyish-buff, with a
marbling of toffee-brown arranged in concentric C-curves
around the outer halves of her eyes. Soft lines of the same
colour shadowed the bases of her protruding eyelids, and a
dash of this "make-up" under the inner bottom corners at
first glance made the eyes seem slightly tilted. Her facial
disc was sharply outlined with a dense, narrow band of tiny,
back-curving chocolate-brown feathers. At top centre this
line curved down and inwards, to be cut by a triangular
dart of brown feathers that grew downwards from her forehead to just above her beak, edged with a V of converging
white stripes.

Tawnies have an acutely hooked beak, to keep it out of
the way of their field of vision, and the thick face feathers
make it look much shorter than it really is. It is made of
bone, but with an outer layer of keratin; Mumble's beak was

dull yellow, fading to pale grey at the sensitive base or cere, where the nostrils were just visible under a fringe of whiskers. Most of the textbooks say that owls have almost no sense of smell, but some experienced handlers dispute this. The fact that tawnies won't touch carrion in the wild, but accept fresh dead chicks in captivity, would suggest that they can certainly tell how old meat is. The whole subject of the sense of smell in birds is only sparsely researched, and much of the work that has been done focuses on a few species for whom it clearly is important, such as the dimsighted, ground-hunting New Zealand kiwi, and some types of roving seabirds. However, a study of the Short-Eared Owl established that the part of its brain that processes smell—the "olfactory bulb"—is of comparable size to that in pigeons and chickens, and bigger than that found in starlings. This is significant, because pigeons, chickens, and starlings are all known to make quite subtle use of their sense of smell. (Dr. Graham Martin, from whose book *Birds by Night* I gratefully harvest this fact, follows it with a dry comment to the effect that it therefore seems "inappropriate to conclude that a sense of smell is irrelevant to owls." And Dr. Martin had a long-lived pet tawny, too.)

Mumble's "whiskers" were a moustache-shaped array of bristly little rictal feathers that spread out sideways from immediately above her beak—the "bridge of her nose"—and extended down and outwards under her eyes. Owls' vision is surprisingly poor at very close range, and they close their eyes to protect them while they are using their beak. There is a rich supply of nerves at the base of these rictal whiskers, so they must trigger sensory receptors that pass information to the brain about anything they touch.

Usually only the central hook of Mumble's beak was visible, above a brown "goatee" mark among the pale feathers covering her throat. However, when she yawned she revealed a wide gape, and the corners of her mouth were actually aligned with the centres of her eyes. The hooked upper mandible overlapped the lower one all round, and both were edged with sharp blades that closed across one another like a pair of scissors. The beak was part of Mumble's killing equipment but also her knife and fork, and to some extent shared the role of fingers with her toes. Since a bird's upper limbs are devoted almost entirely to flight, it has to use its beak to check things out and make small manipulations, and Mumble could use her potentially very powerful bite with great delicacy.

I learned that a bird's tongue has a bone in it, and taste buds at the base (though relatively far fewer than in humans). Even though Mumble often yawned I have no note of ever having seen her tongue; this was shaped like a willow leaf, and she must have kept it pressed down inside her beak. Birds have a larynx, but, unlike that of mammals, it is not used for producing sound. Mumble's voice was produced by an organ called the syrinx, much further down her respiratory tract, and when her beak was closed there was a direct connection between her windpipe and her nostrils—thus, presumably, her apparent ability to "sing with her mouth closed."

Birds have no sweat glands in their skin, and when a tawny is feeling hot or stressed it sometimes seems to pant, with its bill half open and its throat feathers rising and falling. I noticed this in Mumble only at moments of high excitement, such as imminent combat.

◉◉

I KNEW FROM my reading that Mumble's ability to engage acutely with the world around her depended as much upon her ears as upon her eyes, but their extreme refinement was even less visible from the outside.

The Tawny Owl's ear cavities are holes in the lower part of large, slightly banana-shaped trenches set vertically into either side of the skull behind the edges of the facial disc. The textbooks explained that the eardrum is linked to the cochlea, or inner ear—and thence to the auditory nerve—by a single complex bone in the middle ear called the stapes, a relic of birds' evolution from reptiles. This amplifies the sound vibrations by a factor of about sixty-five—three times the efficiency of the three bones of our own middle ear. The membrane in the cochlea of an owl's ear is relatively enormously long, with huge numbers of the hair cells that transmit pressure waves to the nerve that carries the signal to the auditory centre in the brain, so owls can detect much quieter sounds than other birds can. In mainly nocturnal owls like the tawny, this auditory centre is itself several times larger than those of daytime birds of comparable size, such as crows; it is also relatively larger than those of less nocturnal species, such as Eagle Owls and Little Owls. (This obviously makes sense: if you hunt in better light conditions you can usually rely upon your eyes, so you don't need magic ears.)

The owl's brain can register the most minute differences in the timing of the arrival of sounds in each of its widely separated, asymmetric ears—to within about one thirty-thousandth part of a second (about fifteen times better than our ears can manage, despite their being far wider apart). Owls have an innate ability to identify the angle of a sound's origin precisely, to within less than two degrees of the compass. If a sound is repeated, they can then estimate the distance of the source, though this skill seems to take practice to perfect—it is a learned behaviour rather than instinctive, and seems to depend on their familiarity both with the type of sound and with their surroundings. Once acquired, this ability enables an owl to "triangulate"—to accurately judge both the direction and the range of whatever is making the sound.

In conditions so dark that they must rely entirely upon their ears, a first sound for direction-finding, followed by one repetition for range-finding, is enough to guide a Tawny Owl right in. It doesn't make the confident pounce of an attack on something that it can see, but glides in more tentatively, with legs extended and swinging and toes spread widely, to feel for the prey in the place to which its ears have guided it. In one indoor experiment carried out in artificially complete darkness, a mouse was released on a smooth floor with a leaf attached to the end of its tail; the owl's first strike hit the rustling leaf. Importantly, however, owls only seem willing to attempt this sort of blind kill in an area that they know intimately, and when they do so it has been recorded that they then follow an exactly reverse bearing when carrying the prey back to the roost—which must mean that they are relying upon their memory to guide them home.

As with their eyesight, their sound-location is enhanced by the long, flexible neck that allows them to move their head through wide angles in any direction without moving their bodies. They constantly check, update, and compare the stream of signals they are picking up, by bobbing and twisting their heads to vary the position of their ears. Incidentally, there is no evidence at all of owls using "echo-location" like bats—that is, of their making a sound themselves and gauging distance by the delay before its echo is reflected back to them. However, while sharing my flat with Mumble I did notice that she sometimes seemed to mistake the origins of a sound from close range—she might look in quite the wrong direction, even at 180 degrees from the true source. It occurred to me that she might be picking up ricocheting instant sound-reflections from the enclosing walls; perhaps her "computer" was cluttered by these short-range echoes inside a confined space.

The soft crown feathers over most of Mumble's head stood up in a continuous ball-like surface, light brown with a fore-and-aft pattern of dark brown curving lines. I knew that she had feathered flaps of skin along the front and back of the ear trenches just behind the edges of her face, the front flap being larger than the rear one and coinciding with the ruff of stiff little feathers that surrounded the facial disc, but these flaps were completely invisible in the fluffy ball of her head. Occasionally, when I was watching her at rest, I saw subtle movements among the feathers on the sides of her head, and I knew that she was "focusing" these ear flaps to amplify sounds—for instance, raising the front flaps improved her backwards hearing, as if you were cupping your hands in front of your ears.

Once, when we were having a mutual preening session, she rubbed her head sideways against my nose, and the dense feathers parted for a moment. Squinting down, I could see inside her ear: it looked alarmingly big, as if it nearly bisected the side of her skull. It resembled a gently curved trench walled with smooth pinkish-grey plastic, and seemed to have an almost cylindrical "pipe" passing sideways across the bottom, finely webbed with white filaments. (My knowledge of owl anatomy is still distinctly patchy, and I don't think I could bear to attend a dissection. Could this have been the conduit carrying the optical nerves linking her eye to her brain?)

◎◎

IT IS CLEAR that Mumble had wonderfully efficient eyesight at low light levels, and perhaps even more efficient hearing. Nevertheless, in the present state of scientific knowledge it is impossible for us to appreciate the full range of her capability to perceive and make sense of her world. Apart from our very imperfect knowledge about her sense of smell, we can only guess about the ways in which the input she received from all her separate senses and memories was integrated—that is, how it all came together in her brain.

Imagine this: you wake in the middle of the night, and visit the bathroom without turning on a light. How do you do this? (Yes, of course you occasionally—and memorably—stub your toe, but that's usually when your awakening has been sudden, and you start moving before getting yourself mentally centred. When an owl is suddenly startled into night flight, it too may collide with branches.)

Your brain has stored spatial memories of the layout of your home, allowing you to anticipate each step. Your sense of distance covered is imperfect, but your bare feet tell you when you are crossing from one floor surface to another with a slightly different texture. Your eyes are almost useless, but not quite: if the slightest light is seeping in around a curtain, you are aware of dim reflections off pieces of furniture, confirming your mental map. Whether or not you are conscious of it, your ears are picking up sound clues— the quality of echoes from surfaces, the lack of them from empty spaces, the faint sound made by water in central-heating pipes. Perhaps there is even the sound of a dripping tap, and the faintest smell of perfumed soap from the open bathroom door ahead of you (and if the bathroom is tiled, the difference in sound quality will certainly be detectable as you approach it).

Each of your separate senses is providing your brain with an absolute minimum of input, but their integration builds a mental picture of your surroundings that may allow you to move in the darkness with reasonable confidence. If you were an Australian Aborigine, a Kalahari Bushman, or a member of a remote Amazonian tribe—or your own fifty-times-great-grandparent, anywhere in the world—you would still be able to do this out of doors, without consciously thinking about how you were doing it. Well, an owl *can* do it, with laughable ease.

There even seems to be a possibility that owls have access to information from one sense that we do not have. Recent research into other bird species suggests that their ability to orientate themselves by day and night may owe something to a perception of the Earth's magnetic field. Studies of this capability in birds—which have naturally concentrated on migratory species—are in their infancy, but have already

yielded some remarkable suggestions. It has been established that not only pigeons but also some non-migratory
birds, such as chickens, have minute deposits of microscopic
crystals of magnetite (a type of iron oxide) around their eyes
and in their nasal cavities. Separate experiments with
European robins also suggest that a chemical reaction trig-
gered in the left side of the brain by light entering the right
eye is connected with their ability to navigate. The implica-
tion is that these parallel perceptions may be comparable to
a sort of magnetic compass and a magnetic map of the bird's
surroundings. And a significant minority of owl species—
including, in Britain, Long-Eared Owls—do make seasonal
migrations. (Again, I stress my ignorance of ornithology;
but it makes you think, doesn't it?)

◎◎

MUMBLE HAD AN astonishing ability to change her appearance, which was due to the mismatch between her actual body and her loose, elastic suit of feathers. Nowhere
was this more surprising than on her head and neck. The
best way to describe this area of the "outer Mumble" is as an
extremely extendible and compressible toque—one of those
woolly tubes that outdoor types wear around their necks,
which can also be stretched up over the back and top of the
head like a balaclava, while the lower end stays stacked
around the throat in folds. Mumble's "toque" began at the
dark edge of her facial disc, and ended in an invisible merg-
ing with her upper torso. It was a while before she revealed
to me just what she could do with it.

She made it clear from an early age that she hated and

despised pigeons. On the fairly rare occasions when one of these urban scavengers landed on the balcony rail, she would always explode into action straight out of her daytime doze. She would hurl herself at the wire mesh as close as she could get to the intruder, and for some time after the pigeon had exclaimed "Oh, ****!" and fled, she would cling there silently with fanning wings, staring eyes, and half-open beak. When she was free in the flat at weekends I would occasionally hear a sudden flapping and the scrape of claws, and knew that she had spotted one of these (in her estimation) avian guttersnipes getting too close to her territory.

One afternoon Mumble was sitting quietly on the long western windowsill on a favoured perch—a large ceramic keg with a solid top. I happened to point my camera at her at exactly the moment when a pigeon must have arrived on the balcony, and so I saw through the view-finder her transformation, over a period of about five seconds, from fat, contented owl to thin, suspicious owl. It was as marked as the difference in appearance between, say, head lettuce and romaine lettuce; some books say that this response is an automatic attempt by a roosting owl to appear more like a part of the tree trunk it is sitting next to.

First, her head turned towards the intruder, and she fixed her eyes on that bearing with utter intensity. Then, the outline of her body visibly changed: without (on this occasion) extending her legs and "standing up," she compressed her rear shawl feathers so that she grew thinner. Her toque seemed to slip downwards, and a complete change came over her scalp and face. The round ball of her head shrank and altered shape, as the feathers on the back flattened and

Contentment: "I am mistress
of all I survey, and all is
right with the world."

Dawning suspicion: "What's that
on my balcony rail?
Surely it couldn't be . . . ?"

The War Face: "It IS!
It's a ******* PIGEON!"

the frontal crown feathers stood up sharply in a partial Mohawk. The feathers above and between her eyes dropped and stuck forwards horizontally, like bristling eyelashes, and her eyes slitted. Simultaneously, the planes of her facial disc on each side seemed to flatten backwards, giving her a distinctly "hatchet-faced" look. In those few seconds Mumble's whole aspect had changed from benign boredom to unmistakably hostile suspicion. She was suddenly an owl on whose wrong side you definitely would not want to get.

◉◉

WHEN SHE WAS in repose, the feathers on the top and back of Mumble's head seemed to run back seamlessly, in a single hood, down into the thick shawl of light brown scapular feathers covering her shoulders and upper back. Except for the outer edges, which were delineated here and there with white feathers barred with dark brown, it was almost impossible to see where each mottled shawl feather met the next, but there was a slight contrast between the whole mass and the darker contour feathers that clothed her lower back.

Birds have many more touch-receptor cells in their skin than mammals, and it is believed that a layer of hair-like *filoplumes* growing beneath the body feathers gathers information about pressure and vibration. These pass messages to the brain about the relative positions of the various feathers, and thus enabled Mumble to "adjust her clothing" more or less unthinkingly. Consider this ability in human

Mumble "mantling" as she hunts for imaginary mice between the sofa cushions. This "giant moth" pose shows the slightly divided scapular "shawl" on her upper back, and the roughly triangular array of covert feathers that form the uppermost surface of the inner front part of her wing, "thatching in" the bases of the secondaries and primaries.

terms—and be envious. You wake from a nap, and only after you happen to pass a mirror do you realize that you have been walking around with "mad hair." If you were an owl, you wouldn't need a mirror; you would somehow be conscious mentally that your hair was in disarray—and much more acutely conscious of it than your present dim awareness of the pressure of your clothes against your skin. An automatic mental message would pass to muscles in your scalp (admittedly, it would have to be a fairly loose-fitting scalp), and these would shrug your hair back into place and settle it. Who knows? You might even be able to lift a stray forelock out of your eyes without using your hand.

∞

THE WHOLE FRONT of Mumble's body was covered with thick, fluffy creamy-white feathers, each with a central vertical streak of dark brown. (When she moulted, I could see that in fact only the ends were white and brown—the greater part of each feather was made of loose dark grey filaments, but her plumage was so thick that these were completely hidden from the outside.) These insulating layers were deep and very soft, designed to keep owls warm during long winter hours spent motionless on their roosting branches. Birds have a significantly higher body temperature than we do, and a tawny's luxurious clothing is a built-in duvet; when I gently explored Mumble's "wing-pit" with a finger it felt delightfully warm to the touch. These fluffy feathers continued down her belly, and back between her legs, becoming plain white in colour where they covered the bottom cone of her body beneath her tail. Hidden on the surface of this cone, which was an extension of her pelvis, was the single vent that served both sexual and excretory functions (my curiosity never tempted me to take impertinent liberties in this area—Mumble was remarkably patient, but there are limits).

Growing on the separate end of her spinal column, above the cone, were Mumble's twelve long tail feathers or *retrices* ("rudders"), their bases reinforced and "thatched in" above and below by smaller "covert" feathers. The central four rudders were narrow, of almost constant width, and plain donkey-brown apart from a small off-white tip. On each side of these were four broader feathers, patterned with bars like the flight feathers of her wings. When in sustained flight, she spread out all twelve side by side into a broad, rounded fan; as far as I could tell from snatched glimpses, they seemed to be arranged not in a flat plane but in a very shallow arc, with the central four feathers highest and the others slanting away on either side, each overlapping the one below and outside it. When Mumble was settled she swivelled the patterned outer feathers inwards into a single, neatly interlocked stack under the plain central feathers, like the folded vanes of a lady's fan.

◉◉

OWLS HAVE VERY strong legs, with a conventional hip joint at the pelvis and two other major joints (again, a glance at the skeleton drawing on page 167 may be helpful). At the bottom end of the thigh, or femur, the upper of these two joints—once, the reptilian "knee"—flexes backwards, like our knee, but we hardly ever see this under the body feathers. The lower joint, at the bottom end of the now fused tibia and fibula bones of what was once a shin, flexes forwards. This was once the reptilian "ankle"; but below it the reptilian "foot" has elongated enormously and fused into what we perceive as the owl's whole lower leg (the tarso-metatarsus). This terminates in what looks to us as if it should be an ankle, but is actually more akin to the ball of our foot, though with toes extending from it both forwards and backwards.

Most of the time Mumble's skirts of body feathers and

fluff covered her legs down to the second, forwards-flexing joint, and when she settled into her habitual squatting pose the lower legs disappeared too, leaving only her claws showing. She often perched (and slept) standing on one leg only, with the other folded up "into her pocket" among the body feathers. When your instinctive sense of balance is as good as an owl's, there is no point in exposing both your feet to the winter weather at the same time, and thus risking their both being cold and stiff if any sudden emergency or opportunity occurs.

Owls have four toes on each foot, and when they are standing or perched at their ease two of these point forwards and two backwards. However, they have a "trick" joint at the base of the outer rear toe, enabling them to swing it round towards the front at will, to give a three-forwards/one-backwards grip. They may do this in flight when about to land or to strike at prey. Mumble's whole legs and the upper surface of her feet were covered with a fur of fine, pale buff-grey feathers, to keep them warm and to offer slight protection against the nipping teeth of any defiant prey. On the rare occasions when her whole legs were exposed, I noticed that the feathers were thicker around her lower legs and feet than around her upper legs, giving the impression that she was wearing fur boots laced tight at the top and flaring out over the instep below. Among these foot feathers, sensitive filoplumes gave her nervous system a degree of feedback about anything they touched.

The bottoms of Mumble's toes showed pale pinkish-grey skin covered with close-set nodules, and this nubbly "shagreen" surface gave her a secure grip. The talons emerged from high up in big "knuckle" pads at the ends of her toes.

They were glossy dark grey growing from a straw-coloured base, about three-quarters of an inch long, curved like sabres, and needle-pointed. When Mumble stood on a flat surface the upwards curve of the claws from their base in the pads kept all but the points off the ground. When she closed one foot to "put it in her pocket," or both of them in sustained flight, the front and rear talons folded from the knuckle pads right inwards under the toes, like the opposed blades of a jack-knife closing beside one another. I learned that the inner front and rear toes (which do the primary gripping, like our forefinger and thumb) were controlled by a sort of ratchet that allows owls to maintain a crushing grip without consciously exerting constant muscular effort—the grip that enables them to kill their prey almost instantly.

◉◉

I BADLY WANTED to examine Mumble's wings in detail, but—naturally—she refused to tolerate any handling of these delicate miracles of engineering. Her short flights, and her stretches during grooming, were too brief for me to note the details properly, so I had to resort to studying published illustrations, later comparing these with her moulted feathers.

Another glance at the skeleton drawing on page 167 will show how the wing has evolved from the original reptilian "arm" or front leg. From the shoulder joint outwards, it now has three sections that appear to us to be of roughly equal length, though the length of the outer section is almost entirely made up of feathers. The single "upper arm" bone, or humerus, ends in an "elbow"; from this joint the double "lower arm" bones, the ulna and radius, flex forwards, ending in a "wrist"; and from that the partly fused and elongated "fingers," the carpo-metacarpus, flex backwards. (Both the shoulder and the "elbow" joints are invisible to us, hidden by the thick feathers covering the structure.)

Since tawnies are woodland owls, who have to be manoeuvrable to make agile banks and turns during flights between close-set trees, their wings are relatively shorter and broader than those of owls that make more extended flights in the open, like the Barn and Short-Eared Owls. However, for a grown female the wingspan still measures more than a yard, with distinctly separated feathers at the tips. From the tip inwards to about the midpoint of the trailing edge, each of Mumble's wings was furnished with ten of these large, strong pinions or "primary" flight feathers. From the midpoint inwards to the body, the trailing edge continued in a similar row of anything between eleven and nineteen smaller "secondaries" (some sources describe the smaller of these feathers, closest to the body, as "tertiaries").

On the leading edges of her wings, about one-third of the way in from the tip and immediately outside her "wrists," she had separate, outwards-pointing alula feathers, growing from the vestigial "thumb" of the reptilian "hand" that had evolved into the outer one-third of the wing. This bone could be manipulated independently, so the feathers acted like the leading-edge flaps on aircraft wings. (Though in fact, of course, the correct way to describe the resemblance is that the aircraft's flap acts like a very crude and clumsy imitation of the alula.)

Most birds—and some owls—have stiff, glossy flight feathers (*remiges*, or "oars") that cut through the air like knives, but Tawny Owls are among those species that have adapted to sacrifice sheer speed in return for almost noiseless flight. The individual barbules along the leading edges of their primary feathers are not "zipped together" but free, forming a fine comb or fringe. This fringed effect, and a velvety pile over the surface of the feathers, breaks up the turbulence of the air passing over the wings and thus reduces its rushing sound to almost nothing. Coupled with the low wing-loading that makes constant flapping unnecessary, this gives the owl the gliding, virtually silent flight

This side view of Mumble shows to good advantage the various main "panels" of her feathers in repose. The scapular "shawl," edged with white-tipped feathers, stands up a little proud above her folded wings. These are furled and meet behind the base of her tail, crossing at the tips. The shallow arc of tail feathers slants downwards, partially spread.

that is so invaluable when hunting. There is no noise of beating wings to warn its prey, nor to interfere with the incoming higher-frequency sound signals being processed by its sophisticated ear-and-brain computer.

On the top surface of Mumble's wings the background colour of the primaries and secondaries ranged from mid-brown through pale brown to off-white at the trailing edge and tip, and each was barred across with five or six irregular dark brown stripes. The undersurfaces showed the same basic camouflage pattern as the top, but paler, as if misted over with a spray of pale greyish-cream colour.

From the leading edge of the inner part of the wing a series of overlapping rows of velvety brown covert feathers, increasing in size from front to back, "faired in" the bases of the secondaries and primaries.

Mumble usually only opened out the three sections of her wings to full spread when she was either flying or giving them a thorough stretching. The rest of the time she kept the outer two-thirds—from the "elbow" forwards to the "wrist," and from the "wrist" backwards to the "fingertips"—folded up together in a tight inverted V-shape, with the primaries closing underneath the secondaries. From its normal position in repose the thick, smoothly feathered apex of this V looked to human eyes like a shoulder, but it was in fact the bent "wrist," simply held pressed against the side of her invisible shoulder joint. Most of the time she kept what had been the reptilian "upper arm," between shoulder and "elbow," pressed back tightly against her side; she only partially extended it in order to move the two outer sections around as a single thick, closed fan of featherwork.

◎◎

WHILE MUMBLE WAS patrolling around the living-room floor one evening, her vainglorious strut suddenly reminded me of a character in a Japanese samurai film. Like some warrior played by Toshiro Mifune, she had the touchy air of someone who is ready, at an instant, to take furious offence over some imagined slight. She carried her head up and back with her "chin" tucked in, and darted jerky little glances in all directions; once the fancy struck me, I could almost see a tense hand resting on a pair of sword-hilts.

This conceit was followed immediately by another—that her feathers actually had something in common with a sixteenth-century samurai's lamellar armour. In both cases, many small individual elements—for the samurai, laced-together iron strips, and for Mumble, individual feathers—appeared to be assembled into a series of separate larger panels. Like some wealthy daimyo's lacquered, silk-laced composite cuirass, they were intricately barred and powdered with the subtlest of colours. During her first couple of months, when she had still been wearing what looked like a onesie of woolly down, this had seemed to move—if at all—as a single surface. Once she began to come into feather, however, her selective control of her plumage became fascinating to watch.

By muscular action under her skin, Mumble could not only clench or splay the feathers on various parts of her body at will, she could also move whole groups or "armour panels" around independently of other groups. When she ducked to grab at a feather low on the left or right side of her belly and preened it upwards, then that whole half of her front was lifted out of shape in a single mat, while the other side remained flat. This "panelled" effect was very noticeable in the shawl of scapular feathers at the top of her back, which overlaid the rear edges of her folded wings. This shawl usually looked like a single assembly, but occasionally I would see her separating it into left and right halves. She could shrug it far around to either side for preening, and it opened up and lifted out of the way as she spread her wings. When she folded them again, the shawl was puffed out and half extended until the wings were furled tightly against her body under it. Then she gave a little wriggling shake as she dropped it and closed it over the "joins," fairing everything smoothly into a single, apparently continuous surface.

On these and many other occasions, it struck me that my flatmate was not simply beautiful; she was a supremely elegant example of functional natural design.

7

Mumble's Day

THE MOST IMPORTANT daily milestones in Mumble's routine were naturally the simple necessities of life: feeding, digesting the bits that were nutritious, evacuating the waste products that weren't, and grooming herself to keep her feathers in perfect order. Any waking time left over from these essential tasks was spent in keeping a beady eye on her surroundings, and—particularly in her younger years—in playing "war games."

◎◎

EVER SINCE SHE arrived in the flat she had been able to swallow a chick whole, so I was able to forget the repellent task of scissoring them in half that I had needed to perform for Wellington. Since a rough estimate of her body weight suggested that she needed about four ounces of food daily, the basic ration-pack was two chicks. (This was not invariable, because even the European Union has not yet enforced regulations demanding that chicks come in perfectly standard sizes.) Usually I fed her these last thing at night, but later I slipped into the habit of dividing them between supper and breakfast. My notebooks don't explain this; it may have been her idea, but I suspect that it was so that I had a means of tempting her to get into the basket for transit to the balcony cage on the mornings when I had a commuter train to catch and she was playing hard to get. (At the time I never gave any further thought to this, but much later

I realized that I must have been duplicating the time she spent on digestion each day. Luckily this seemed to do her no harm, since she sometimes simply chose to leave part of a chick on one side as a snack for later, but it played hell with my already doomed attempts to calculate when—and therefore where—she would defecate.)

On our first few nights together I had tossed the supper-time ration into her kitchen cage as a way of luring her in there for the nightly lock-up, but this very quickly became unnecessary. As soon as I whistled, or took thawed chicks out of the fridge and dunked them in hot water to take the chill off, she flew straight into the cage and sat impatiently waiting for me to serve them up, often with a little chitter of eagerness. If I offered a breakfast chick when she was already in the balcony cage, she leaned forwards so impatiently and so far that she had to spread and flap her wings for balance, half-hovering at an angle of 45 degrees but with her feet still clutching the perch. In this gargoyle posture, she snatched the chick with her beak, making metallic chitterings with her mouth full as she wound her body back through the air to a stable base. Then she transferred the chick to one foot and held it in her talons for a while, looking around, before jumping down to her shelf and bending her head to start feeding. If she happened to let it slip and it fell to the floor of the cage, she dropped like a stone and stood guard over it, mantling her wings above it protectively and glaring fiercely around before seizing it again and carrying it back to the perch.

Mumble didn't object to my watching her eating, being wholly absorbed in the business of holding the chick down with one foot while dipping her head repeatedly to tear chunks off it. After each bite she would raise her head again with her face upturned, to straighten her throat and ease the swallowing process. She seemed to close her eyes as she bent down, and half-opened them to slits as she reared up to swallow her mouthful down with a couple of big gulps. She often left the legs until last, and the occasional sight of her eating them from the thick end was mildly disturbing— she ended up with a little, vaguely humanoid "hand," or even a pair of them, sticking out of the corners of her bill until they disappeared with a final couple of gulps. When she had finished eating she usually "feeked"—stropped both sides of her beak against a perch, presumably to clean it of drying blood and yolk. (She also often used to bite and rub at her perches between meals, presumably out of an instinct to keep her beak honed down; if the upper mandible grows into too long a hook it may impale the food awkwardly. The pinewood edge of Mumble's big tray-perch was a favourite whetting-stone.)

Mumble's interest in her food even seemed to extend to the logistic side of things. On one occasion she was free in the kitchen while I was reloading the freezer cabinet with day-old chicks from a sack. She paid very close attention, sitting at first on my shoulder and watching with what seemed to be rapt interest as I counted them into plastic bags. When I opened the freezer door she jumped to the top edge of it, head between her feet, watching the passage of each bag as I stacked them inside (I had a mental picture of her holding a clipboard and pen, ticking the bags off on a loading manifest). When I finished stacking the cabinet with a "rock face" of plastic-wrapped chicks, she even tried to get inside with them, clinging onto the icy ledge at the bottom and flapping her wings for balance.

Mumble on her usual perch on top of the living-room door, to
which she flew back "with the beetle buzzing in her clenched
foot, to crunch it at her ease." As she bends her head to nibble at
the treat that she is holding between her toes, her eyes have
closed to protective slits.

You may hear it said that birds of prey do not drink, getting all the moisture they need from their food, but nobody
who holds that opinion has ever shared quarters with an owl.
When Mumble was out in her balcony cage I would quite
often catch sight of her sitting on the edge of her water dish

and dipping her head down between her toes. She would delicately scoop up sips of water and then tip her head back to let them slip down, visibly swallowing, with her throat working and eyes blinking. One evening in her second year I walked into the kitchen to discover her sitting on the edge of the washing-up bowl in the sink and moving her head around under a dripping tap, letting the drops fall gently into her open beak. After this I would sometimes deliberately leave the tap dripping for her, with a sponge placed below it to stop the noise driving me crazy.

◎◎

NO HONEST ACCOUNT of life with an owl can ignore what we might delicately call "the disgusting bits."

Unlike daytime raptors, owls have no crop or "storage cupboard" in their throat, but a two-stage stomach. Since birds have no teeth and cannot chew their food, they swallow it whole, and this means that fledgelings have to learn to tear their prey into swallowable chunks. While Mumble managed her habitual chicken dinners efficiently, in later years after we moved to the countryside she was sometimes over-ambitious with the unfamiliar prey that she occasionally caught for herself "on the hoof." The sign that she might be regretting her greed was an apparently uncomfortable immobility, sitting ramrod-straight on her perch with eyes slitted and upturned face stretched tight, and somebody's tail protruding from her half-open beak.

Once successfully swallowed, the food passes into the "pre-stomach," where powerful acids and enzymes break it down. As already mentioned, to power its demanding lifestyle an owl needs roughly 20 per cent of its body weight in food every day, so the digestion process has to be relatively fast and simple to clear the way for the next meal. The hard or indissoluble parts—bones, teeth, beaks, insect wing-cases, fur, and feathers—then pass down into a gizzard or "lower stomach," where, to save time, they are formed into pellets for regurgitation. (It is sometimes assumed that only raptors produce pellets, but in fact hundreds of species of birds do this—in Britain, for instance, these include not only kingfishers and herons, but also rooks, starlings, tree sparrows, and even robins.) Meanwhile, the useful bits are absorbed for nutrition, and—inevitably—that produces waste that has to be excreted.

Since birds do not have a separate bladder for urine (one of the many weight-saving modifications to their innards), an owl's "unified" droppings are a strongly acidic, foul-smelling brown-and-white sludge. This is expelled backwards horizontally, with some force—a procedure known to falconers as "slicing," which resembles a tobacco-chewer spitting out a thick stream of evil brown juice. Anyone carrying a bird on their fist quickly learns to keep half an eye at all times on the direction in which its bum is pointing—and particularly when innocent civilians step up close to admire it. Since you get only between two and three seconds' warning, it's important to be alert for the signs of imminent action: a slight crouch, a thoughtful expression, followed immediately by tails-up, a parting of the fluff, and—"*Torpedo . . . Los!*"

I used to try in vain to work out at roughly what times of day, in relation to her feeds, Mumble was likely to slice, but it was hopeless. I was slightly more successful in figuring out the probable danger areas around the flat, laying copious newspapers on the floor and spreading thin plastic sheets around. The most obvious ground zero was a radius around her favourite perches, but she sometimes fooled me with her insouciant timing, and I simply had to accept that stains, scrubbing, and subsequent patches of bleaching on the living-room carpet were inevitable. (Luckily, it was cheap, and I had never liked the colour anyway.) My willingness to pay this price for living with an owl was probably the eccentricity that my urban friends found hardest to understand.

By contrast with this unavoidably unpleasant business, the pelleting (so useful to scientists researching owl diets and distribution) was a sedate affair. In the gizzard the indigestible parts of the prey were neatly formed into a short, sausage-shaped parcel wrapped in the tightly packed debris of fur or feather, which passed back up to the pre-stomach. There it remained for several hours before being regurgitated. In the wild this usually happens while the owl is roosting during the day, and the textbooks say that during the digestive process no new prey can be swallowed—a fact that I was slow to pick up on, so (as mentioned) Mumble often got her rations divided between supper and breakfast. Nevertheless, she seemed perfectly able to shuffle it all around internally without discomfort.

The signal that Mumble was contemplating bringing up a pellet began with her yawning widely; then she would bow low and shake her head vigorously from side to side at a rate of four or five shakes per second. She would then pause, sit upright for a couple of seconds, and then bow and give another rapid series of head-shakes. She repeated this whole sequence perhaps four or five times, and then gave a series of huge yawns—every ten seconds or so, for up to a minute at a time. After a further series of head-shakes in an upright stance, she might forget the whole thing, as if deciding that it had lost its urgency for now—like a human misled by an apparently imminent sneeze that stubbornly refuses to happen. But if she was ready, then she would bow again, close her eyes, shake her head—and the pellet would drop neatly out of her beak. There was no gagging or spitting reflex; the pellet would not appear during a yawn, but during one of her sideways head-shakes. Given her almost exclusive diet of chicks, the smooth pellets were yellowish-grey; the initial slick of moisture dried off fast, and there was nothing at all distasteful about handling them.

This process did not seem to require much concentration or to make Mumble feel vulnerable or self-conscious. If I happened to walk past while she was yawning or shaking her head, she might occasionally hop down to my shoulder and carry on there, as if nothing remarkable was happening.

◎◎

MUMBLE CLEARLY GOT a great deal of sensual pleasure from stretching, and when I caught her at it I found myself mouthing ecstatic creaking noises in empathy.

Standing on a perch that gave her a good deal of clear space behind her feet, she would balance on one leg and lean slightly towards that side. Then she would slowly and deliberately stretch the other wing out and downwards, her primary and secondary feathers spreading like fingers, while simultaneously stretching that leg and foot downwards inside the wing, with the toes splayed apart. After holding

this pose for a couple of seconds, she would gently furl up
the wing, return that foot to the perch, lean to the other
side, and repeat the performance with her other wing and
leg. When she was centred once more, she would crouch
forwards with her head over her toes, and crook both wings
stiffly up above her back in symmetrical L-shapes bent at
the "wrist," as if she was imitating the eagle standard of a
Roman legion. Finally, she would stand upright again, shrug
her refolded wings into place comfortably, and give her
body feathers a settling shake. She was once doing a wing-
stretch as I walked through the door below her; simultaneously, she squeaked at me, continued the stretch, and was
gripped by a sudden urge to yawn cavernously. This made
her look rather like a cartoon opera singer making a hugely
theatrical gesture as she reached her top note (though
Mumble's came out like the squawk of a tin trumpet).

STRETCHING SEEMED TO be a casual and unthinking self-
indulgence; by contrast, a full-scale grooming session was a
wholly absorbing business that could easily occupy Mumble
for up to an hour of methodical maintenance work.

All birds groom themselves at frequent intervals, both to
clean their dense plumage of dust and parasites, and to keep
their flying surfaces zipped together. The individual barbs
of their flight feathers are hooked together into a continuous vane by means of tiny barbules along the edge. These
come apart with vigorous use, and the bird "re-zips" them
by running the feather through its beak. Mumble usually
started the grooming session with her wings, lifting them

and twisting them inwards and forwards as she selected frayed or disordered flight feathers one by one and dragged them through her beak slowly from side to side, making her machine-gunning noise at about eight beats a second as she zipped their edges tight again. Sometimes I caught her doing this "backwards," with her head twisted round and half-way down her back while she pecked at one of her primaries.

As a spectator, I found Mumble's body-grooming a never-failing source of entertainment—she was so impossibly ag-ile, and just so damn *busy*. During these sessions her head was never still for more than a few seconds. With eyes closed or slitted, it nodded up and down, twisted round, and tilted from side to side as she rummaged and nibbled with her beak among the roots of her feathers. Her suit of skin gave the odd impression of being oversized, loose, and attached at only a few essential points, and her body-grooming demonstrated yet again that her plumage was grouped into independently movable panels. She could shrug the rear shawl feathers to left or right, bringing almost the whole group around one shoulder in either direction to within reach of her beak. To reach her throat feathers in-volved making her head appear much smaller and flatter-backed than usual. The toque of head and neck feathers slipped down into a bulky ruff, which she nibbled by push-ing her compressed head back and tipping her face for-wards, catching the ruff under her "chin" and then working her beak slowly round to one shoulder, rootling hard and peck-peck-pecking.

Tawnies are physically capable of rotating their heads a full 270 degrees. That is, in human terms, as if you started with your face forwards (twelve o'clock), then rotated it

180 degrees to the right until you were looking straight backwards (six o'clock)—then continued to turn it to the right until you were looking in line with your left shoulder (nine o'clock). Even if your neck had as many vertebrae as an owl you couldn't do this without losing consciousness, because the compression of the major blood vessels as your neck twisted would starve your brain of oxygen; owls have a special by-pass adaptation to their carotid and vertebral arteries that prevents this happening. (In practice, of course, Mumble usually had enough sense not to twist her head through more than 180 degrees—if she wanted to nibble at one side of herself, why go the long way round?)

When she hunched her head down into her toque-feathers with no visible neck and bowed face-first in any direction to reach her lower parts with her beak, the stretching extension of her long neck was completely hidden inside what looked like a single, continuous ball of feathers incorporating head and body together. When she closed her eyes and rotated her head so that she could push her beak down into some part of the fluffed-up mass, her face seemed to disappear while it sneaked downwards in some unpredictable direction. When she half-opened her eyes again in mid-groom her face suddenly reappeared, apparently growing out of an impossible part of her anatomy at an unexpected angle. The difficulty of telling, at any particular moment, which bit of the mass of feathers was Mumble's head, and which wasn't, was aggravated when she pushed her upside-down face backwards far under a wing, nibbling at the "wing-pit" from underneath. When she turned her head 180 degrees backwards and ducked it down to peck at the base of her tail, then from the front she appeared to be completely headless. When she groomed her tail feathers she could manipulate their angle sharply in any direction, lifting and spreading them like fingers so as to get beak-access to individual quills.

After I had watched this irresistible contortionist cabaret a few times, I began wondering why she seemed to spend so much time with her face rubbing and rolling just above her tail. Watching her from behind, I then got a momentary shock. The thick, dark feathers above her tail parted, and I saw a kind of pink pyramid protruding from the greyish skin in a little clearing between the feather shafts. For a second I wondered in alarm if this was some kind of injury, but then I saw Mumble's "disembodied" face sliding down through the thicket of feathers towards it. She began stropping the side of her beak against it with every sign of satisfaction, before straightening up and shrugging the feathers

The Headless Owl—Mumble seen from the front, while preening the small of her back.

closed to hide it again. Ridiculously, I felt momentarily embarrassed at having intruded on what seemed like a particularly intimate phase of her toilette. A return to the textbooks taught me that this pink spike was her uropygial or "preen" gland; by rubbing it she stimulated the release of an oily liquid which she then rubbed over her feathers—both to condition them and to stimulate the production of vitamin D when she sat in the sunshine. (Incidentally, I was surprised to learn that while Barn Owls have this gland, it produces no oil.)

The only part of Mumble's body that she could not reach with her beak was her head itself, and as part of her grooming routine she would give the whole feathery ball a thorough raking with the claws of her free foot, like a dog scratching behind its ear with one hind leg on auto-rotate. She would set up a vigorous back-and-forth scratching with the razor-sharp points while delicately turning the facets of her head and face one way and another against this blurred buzz-saw, as tiny bits of feather-trash flew in all directions. When she stopped, all her face feathers would be standing out in a furry mass for several seconds before they gently sank back into place. (Again, on one occasion the urge to give her face a good scratching coincided with the pelleting process. She was sitting on a perch, yawning hugely; then she tilted her head sideways for a vigorous scratch, yawned again in the middle of it—and fell off into mid-air in an undignified bundle, her balance on one leg obviously unequal to the satisfaction of both urges at once.)

She ended any grooming session with a huge fluff-out and a hard, clearly audible shake, five or six times back and forth, with all the body feathers rising and separating. This was followed with a last prim, Victorian little shrug to settle the edges of her furled wings under the shawl and the fluffy edges of her breast feathers, and a final shuffle of her feet.

◎◎

MUMBLE'S MOST CONSTANT occupation, from which all other pastimes were temporary distractions, was simply observing. Her job, her hobby, her passion was watching things—which is hardly surprising, given her place in the natural order. From her various preferred vantage points around the flat she kept constant surveillance over her environment, and when she detected any hint of sound or movement her evaluation of it was presumably based on the central question of any carnivore's existence: Can I jump on it, or is it going to jump on me?

If it was small and mobile (which, in a seventh-floor flat, meant it was an insect), then it could travel only a matter of inches before Mumble arrived like a Stuka; but if it sat still, being enigmatic, then she might settle down to out-stare it. This balancing of boldness and caution in the face of the unrecognized cannot have been learned in her own short, protected lifetime, so it was clearly part of her genetic inheritance. Down tens of millions of years, was *Protostrix* still whispering a race-memory about dead twigs that turned out to have teeth? If so, Mumble was soon reassured that in this flat she was the pinnacle of the food chain. Her problem was that she was so much larger than any potential live prey that simply picking them up was tricky.

During her second summer, she noticed flies sitting on the ceiling of the living-room. This provoked a series of

A companionable breakfast-time on a sunny Sunday
morning in the flat; I'm evidently reading a newspaper,
into which she may very well plunge without warning
at any moment, kicking and flapping for the sheer
hell of it. In the background is the bust of
Germanicus Caesar, a favourite perch.

determined but unsuccessful attacks. She sat on all possible
perches glaring up at them, bobbing and weaving to follow
their flights until it looked as if she would screw her head
right off. When at last they seemed to have settled, she
crouched, leapt into the air, and climbed in a frantic flurry
of wing-beats. At the last moment she did an impressive
aerial back-flip, belly parallel to the ceiling, and struck upwards at the flies with her feet. Naturally, she missed: they
slipped out between her toes, leaving her to stall away in
inverted flight, flapping madly to regain control before she
hit the carpet. She would then repeat the whole pointless effort several times, until she got fed up and flew off in a sulk.

On one occasion (or one that I saw) she spotted something big and slow enough to give her a reasonable chance. Her sudden fixed stare drew my attention across the room to where a small beetle was climbing up the curtains. Feeling that she needed some encouragement after the fly debacle, I picked her up on my fist and carried her over, holding her close to the beetle. *Peck!*—near miss, disturbing the curtain; the beetle fell a few inches, but hung on and resumed its climb. *Peck, peck!*—another near miss, and this time the beetle spread its wings and flew off. Together, in a parody of hawking, we stalked it until it landed on the ceiling. Mumble watched it with laser intensity, settled herself, then kicked off mightily from my fist. A perfectly timed back-flip and upwards strike, an elegant Immelmann turn— and she was flying off with the beetle buzzing in her clenched foot, to crunch it at her ease on the door top.

◎◎

IN THE RELATIVE absence of live prey, Mumble made do with war-games. She had learned early on the joys of jumping down from a height into a wastepaper basket and doing a war-dance among the satisfactorily scrunchy contents, but her main pleasures were pursued at higher altitudes.

One of the ceiling-light fixtures in the flat was fitted with a "Chinese lantern" shade—a big white paper globe formed around spiralling circles of thin wire. Mumble soon discovered that attacking this was fun; the paper tore under her claw-kicks in a pleasurable way, while the globe bobbed about excitingly, making *bonk, bonk* noises. The climax of this First World War "balloon-busting" game occurred when she had punched so many holes into its shoulders that its increasing fragility and the slight weight of the wire armature reached a critical tipping-point. One evening a final attack was enough to tear apart the shreds of paper remaining between two of the worst stretches of battle-damage. This started a chain reaction, and as Mumble fluttered excitedly around it the whole spiral assembly slowly unravelled downwards towards the floor, like an apple peeling itself.

She was crazy about table lamps, and during our first year she cost me several broken ones by landing hard and clumsily on the upper edge of the shade, knocking them to the floor. To my surprise, the bright light shining in her eyes did not seem to bother her at all. If a lamp survived one of her brutal landings still upright, she would cling precariously to the thin rim of the shade, wings half-open for balance, chin on her toes, gazing right down into the brightness of the bulb. The attraction sometimes seemed to be the warmth; she would close her eyes and wobble on this insecure perch for minutes at a time, apparently basking. During one of these sunlamp sessions she got a bit too dopey, and actually jumped deliberately down into the shade, with her feet on the top of the bulb. Discovering too late that it was burning hot, she found herself confined within the narrow funnel of the shade with her wings extended above her head. She gave a convulsive heave of the legs and exploded upwards, while I grabbed vainly for the falling lamp. Before long I learned to buy much heavier-based replacements.

◎◎

MUMBLE WAS EVEN more certain death to houseplants. Sometimes she dropped heavily amongst the greenery from above, kicking and pecking, but the best game was stripping them on the wing. With their mass of long, hanging fronds, spider-plants were her favourite prey. She would make a carefully calculated swoop across the room, snatch one of the leaves off as she passed in a fast, climbing turn, and carry it off to her door top. There she would seem to try to eat it, although this was completely against nature; she would hold the frond in one clenched foot like a stalk of asparagus, biting bits off it and tossing her head back as if she was trying to gulp them down. However, she always ended up by dropping the shreds over the edge and watching with rapt attention as they fell to the floor. I could follow the steady destruction of my houseplants by the faint pattering of debris falling on the newspapers spread below her perches.

On one occasion she engaged in a ludicrously show-offy variation of this game. Instead of bearing her catch off to a familiar perch, she flew to the alcove where I had hung a print of Albrecht Dürer's charming 1508 study of a young tawny. There she posed, clinging to the deep frame as it tilted under her weight, with the frond clamped across her beak. When I went across to lift her off she bugled triumphantly (speaking with her mouth full always gave her voice a strangely metallic ring).

When Mumble first moved in, the biggest spider-plant was growing in a pot on a wooden pedestal that stood in a corner at one end of the living-room window-wall. After she had knocked the pot over twice during her skip-bombing games of *633 Squadron*, I got fed up with sweeping compost

The episode when she bore her trophy up to the
frame of a print of Dürer's drawing of a young
Tawny Owl, clinging on and bugling
triumphantly as the frame lurched beneath
her weight.

off the carpet. I moved the plant pot to the end of the west
windowsill, guarding it from further plucking by the strate-
gic placement of the curtain and of a big, vicious Moroccan
crown-of-thorns plant that she had enough sense to give a
wide berth. Rather than leaving the pedestal pointlessly
empty, I replaced the plant pot with a sturdy near-life-sized
bust of Germanicus Caesar (elder brother of Claudius,

Although she would never actually eat any
plant matter, she loved snatching fronds off
house-plants and bearing them off to a perch,
where she would hold them like stalks of
asparagus and slowly tear them to pieces
with her beak.

and—to judge from the writings of Tacitus—the best man
in a spectacularly dysfunctional family).

Mumble quickly took to this change in the interior dé-
cor, and thereafter she would roost happily on the general's
sculpted poll for hours at a time. I thought this gave the
room rather a classy, Edgar Allan Poe–ish look, and luckily
Germanicus scrubbed clean easily. For when she wanted a
view of the sunset, I gave her a large ceramic barrel to perch

on in the middle of the west windowsill, well away from the crown-of-thorns. This too seemed to suit her, and thereafter she divided most of her living-room time between the door top, Germanicus, and her barrel.

◉◉

BY THE TIME we had been together for a couple of years, Mumble and I had long settled into a generally relaxed and companionable routine.

When I got home at night she heard my key in the door, and I would hear a long, liquid warbling from the balcony. By the time I got out there she was always inside her hutch, with her head shoved in a corner, whooping away for several minutes before emerging onto her doorstep perch. When she had finished her waking-up routine and I went inside the cage, she would sometimes come to my shoulder to greet me (and when she pecked my moustache on chilly nights her beak smelt a little like cold, wet beach-stones—there was no foul carnivore's breath, like a dog's). On other evenings she would jump directly from her doorstep to the open basket in a single accurate hop, avoiding the other perch in her path. When I got home after leaving her free around the flat she would come flying straight to my shoulder, sometimes warbling or crooning. Since she was by nature a solitary creature, and entirely accustomed to being alone for most of her life, this was not just a sign that she was pleased to see anybody—she knew me from other people. Since her second summer she had reacted to them with unfailing hostility, and if she was free to do so she attacked them; so, her greeting was specifically for me.

When it was the other way around, she hardly ever tried to evade my touch. When she was perched on Germanicus and I walked close by, I sometimes could not resist gently stroking her head, or bending to nuzzle her between the eyes or on her downy front. Sometimes she accepted these caresses complacently; sometimes she would squeak softly, shifting her feet and giving me a very gentle and half-hearted peck on the nose; but she never made to claw at me, or to fly off.

In fact, she never, ever attacked me with her talons, even when struggling angrily not to be put in her basket for transfer to the balcony cage when I had to leave for work in the morning before she had decided she was ready. On these occasions I sometimes had to lift her off five or six different perches in succession. The correct way to pick up a bird is to run your hand up the back of its legs, and it will naturally step up backwards onto it. If I was determined to dislodge Mumble from a position where this was not possible, and ran my hand up the front edge of a perch and under her forwards-projecting claws, she regarded it as a gross breach of good manners, but it did work—for a moment she was too busy regaining her balance to do a mid-air take-off. She might be pretty furious by the time I finally grabbed her; she would chitter with rage and peck at my hand, but only gently, and she never used her claws on me. This seemed to me to be remarkably *measured* behaviour, and surprising.

(The only time she ever drew my blood was entirely my own fault. I stood up carelessly in her path just as she was flying one of her fast, silent evening runs down the length of the flat, and although she jinked in mid-air to avoid me, a trailing claw opened up an inch of my cheekbone like a razorblade.)

On the other hand, she often sought my touch deliberately, and not just by flying to my shoulder. When she was there, she sometimes ducked her head down and preened the side of my beard, making her little machine-gunning noise with her beak. On one memorable occasion when I had left the night cage door open after feeding her, intending to leave her free during the night, she suddenly appeared on my shoulder with a disgusting tangle of partly eaten chick in her beak. She repeatedly leaned around my face, unquestionably trying to reach my mouth and feed me. When I avoided these deeply touching but unwelcome attempts, she tried a couple of times to stuff the slimy gobbet into my ear instead (I'm not making this up).

Her demands for "heavy petting" sessions had grown less frequent since her first year, but they still happened every few days. When she was in the mood, once or twice during an evening she would walk across the floor to my armchair and climb onto my lap. When she was settled she craned her head upwards, cheeping quietly with her eyes closed, until I dropped my face and nuzzled her. When I did this she twisted and pushed her head against my nose, like a cat. The fact that she still did this regularly gave me as much pleasure as it seemed to give her.

☉☉

IN LATER YEARS, after we had moved to Sussex, I noted a very definite correlation between Mumble's demands for my "preening" and the summer moulting season. Yet, puzzlingly, when looking back through my notebooks I find no clear mention in the years 1979–81 of any noticeable moult. I know that I found numbers of fluffy body-feathers and some small quill-feathers lying around the balcony cage at times, but there is no note of her losing primary or secondary flight feathers. I cannot imagine that I would have missed this, nor the very "clingy" mood that accompanied the process in later years.

The textbooks say that owls have a partial moult of body feathers and some wing coverts in their "post-juvenile" phase. They then change all their tail and some flight feathers for the first time in their second summer, after having fledged their first clutch of nestlings. (It seems unlikely that the act of raising a brood might actually be the trigger for the moulting process. Even if Mumble's virgin state delayed its first onset, from 1982—her fourth year—her moult was both heavy and reliably annual. I have since read that owls in captivity often don't moult properly until they are at least two years old.) Tawnies don't lose all their flight feathers each year, only about one-third of them, so a bird doesn't acquire full adult plumage until its third autumn. Consequently, a mature bird's wings may show feathers from three different years, and experts can sometimes tell these apart by small differences in pattern and fading.

◎◎

EARLY IN 1981, I began to notice that Mumble was getting rather less agile and energetic than in the past. Her usual daily rhythm had always been essentially that of a cat—that is, she spent about twenty hours dozing, an hour or two on self-maintenance, and two or three hours in alert activity and exercise. Now she seemed to be getting even more lackadaisical, and I wondered uneasily if she might have some health problem. (I had never had to try to find a vet in South London who understood owls, for which I was grateful. Having decided from the first that I would never tether her legs, it would have been a hilariously difficult and probably bloody task for a third party to handle her for examination.)

A simpler reason for her new languor suggested itself one night when I was ready to feed her and put her to bed. She had been sitting on her barrel on the end windowsill, at the full length of the flat from the kitchen cage, when I got out her chick and whistled for her. She did not appear instantly, so I wandered to the kitchen door to look down the living-room. She was still on the windowsill when I whistled again. She took off, flew half-way down the room—and then stopped to sit on the back of a chair. I may have imagined it, but her body language suggested that she might actually be wheezing . . .

All became instantly clear. The reason that this owl was now lurching through the air for a pathetic few feet instead of flying like the Red Baron was that this owl was getting shamefully fat. Birds of prey operate best if they are permanently aware of a small corner inside them that could still do with filling, even after they have swallowed the last beakful and mopped up the gravy. Unlike a falconer with a working bird, who must weigh it daily to calculate the optimum level of slight hunger for efficient hunting, I had hardly ever bothered to persuade Mumble to sit on the kitchen scales, and even when I did she wriggled too much for more than an approximate reading.

She had become a feathered gourmande, and had been living too soft for too long. Obesity is a problem that is never, ever encountered by raptors in a state of nature. Mumble had been designed as a superlatively effective night-fighter; now, thanks to my soppy indulgence in the matter of an extra chick pretty much whenever she demanded one, she would soon require the take-off run of an overloaded 747 on a tropical mountain airstrip. It occurred to me that if we kept up this regime she might soon give up flying altogether and become the first-ever wholly and shamefully pedestrian Tawny Owl. Since I was doubtful of my ability to persuade her of the virtues of muesli and jogging, the only solution was a strict diet. I have never believed all this pseudo-scientific hogwash about the female brain having a special mystical need for chocolate; greed is greed, and for at least the rest of the spring and summer Mumble was going back to two chicks per day.

◎◎

ONE MUGGY FRIDAY evening in the spring of 1981, I got off a crowded commuter train from Victoria Station at about 7:30 p.m. and began the nightly ten-minute walk down the main street to my building. I walked automatically, with my mind miles away somewhere inside my head. I was tired, and grubby; it had been one of those city days when you can positively feel the sweaty grime building up inside your shirt collar. I had no plans for the evening, and my general mood could be summed up as *bluuuh*.

It was a straight walk, punctuated only by crossing the ends of various side-streets where they joined the London road. I was half-way through my journey when my auto-pilot tripped off and I noticed something unusual. It was early on the first evening of a warm weekend, in the last big urban centre on this edge of South London—yet there didn't seem to be much traffic passing, nor many people on the sidewalks. As I crossed one of the side-streets I automatically glanced right to check for cars coming, and saw a group of youths hanging around just back from the street junction. There were about eight of them; they weren't doing any-thing, just smoking, drinking from beer cans, and talking quietly, standing in a tight knot. They seemed to be waiting for something to happen. I kept walking, and a couple of hundred yards further on, across the street in the mouth of another sideroad, I saw a police van parked, with half a dozen officers also standing around expectantly. On any weekend the town-centre pubs and the dancehall-cum-nightclub next door to my building were always lively, but this seemed different—what did all these people know that I didn't, on this oppressive Friday evening? I went straight up to my flat, and stayed there.

During an evening that was punctuated by the sounds of shouting and sirens, I did a bit of thinking about my life. I lived in the sky, but when I came down to ground level it was into an environment of dirty concrete, diesel fumes, and crowded sidewalks. I travelled to work in cattle-truck comfort, arriving among more concrete, thicker fumes, and even tighter crowds. True, I passed my days surrounded by all the attractions of one of the world's great cities—but most of those days were spent shut up in an office. When I slogged home again it was only to take refuge from the same old ugliness, stink, and noise, and although office and flat were not many miles apart as the crow flies, the journey door-to-door usually took an hour and a half. Not for the first time, I asked myself if I really wanted to do this for the next thirty years. Over the weekend, I came to a conclusion: like Huck Finn, it was time for me to "light out for the Territory."

◎◎

SELLING MY FLAT would be no problem; the housing market was buoyant, and the easy access to central London would be a positive attraction for somebody less jaded than I was. But where should I look for a new home? I had been born and brought up in a pleasant-enough village outside a boring commuter town in Surrey, but since my parents died I had had no further ties in that direction. On the other hand, there were good train connections between London and the Sussex coastal towns, and I had always loved the South Downs country. (A phrase from a thriller novel had stuck in my memory as particularly apposite: the villain had been described as leering at a damsel in distress "like a Norman looking at Sussex.") Most of the towns along the coast had the reputation of being waiting-rooms for Heaven, but Brighton was always lively—"London-on-Sea." It was also half-way between the homes of my brother in Kent and my sister in Hampshire, an easy ninety minutes' drive to the east and west respectively. I took a map, stuck a compass-point in Brighton, and drew a ten-mile circle around it; then I hit the estate agents.

You know how these things always go. Every Saturday noon for the next couple of months I drove off southwards with the passenger seat covered with estate agents' paper-work and a schedule of viewing appointments; and every Sunday evening I drove home again with the floor-well full of torn-up paper, in a mood of seething frustration. Predictably, at exactly the point when I had become despondently convinced that there was no single house in Sussex that I could both afford and bear the thought of living in, I found it. It was described on the last page of that weekend's sheaf of extravagant lies; it was an unremarkable three-bedroom duplex built in the early 1960s, and it sounded nothing special. But it would only mean a detour of ten miles on my way back to London, so after a mental toss-up I decided I might as well give it a drive-by glance.

As I navigated the last few miles I began to feel a faint stirring of interest. I found that this address would be within forty minutes' drive of my old friends the Hook family. The ancient town nearby seemed charming, with a half-ruined Norman castle at the top of a steep, busy hill down to a river bridge. The village itself looked as if it was a work-ing community rather than a dormitory—I had to over-take a tractor with a laden trailer as I drove past a pub facing the cricket green. There was another pub right at the end of the lane I was looking for, and when I turned up this lane I found myself driving between field hedges as well as stretches of housing. The house for sale was almost at the far end, just before fields took over entirely.

I parked up, and took an oh-so-casual stroll past. Natu-rally, the estate agent had not bothered to mention the at-tractive location. I couldn't see into the back garden, but a hundred yards further on there was a footpath leading off in that direction, so I took it. I found myself climbing over a stile into an open field, and walking up along a hedgerow. The garden of "my" house was of reasonable size, and was shaded by a noble oak tree; I could see that the view from the back windows would reach westwards over fields all the way to a line of green hills. As the Iron Duke is supposed to have exclaimed at the key moment during the Battle of Salamanca: "Damn me—that'll do!"

◉◎

I IMMEDIATELY MADE an appointment to view, and found that the owners—a scientist at Sussex University and his wife—were pleasant and reasonable. The inside of the house was as satisfactory as the outside, the whole property looked owl-friendly, and—hallelujah!—the long living-room even had an open, working fireplace. The asking price was realistic; it was at the top end of my range, but manageable. I found a buyer for my flat without difficulty; then all three corners of the deal entered that unavoidable limbo period during which solicitors and mortgage-brokers torture everyone concerned until you are all on the screaming edge of a nervous breakdown. Eventually, all the contracts were exchanged and money changed hands.

I had a couple of weeks' grace between the previous owners' moving out and my moving in, so I could make several car journeys with much of my more awkward gear before the final trip with the removal van. I also got my brother and brother-in-law to meet me there one weekend and help me construct Mumble's new country quarters. In a leafy corner at the bottom of the garden Dick, Peter, and I built a sturdy timber and wire mesh aviary rather larger than the size of my own bedroom in the flat, complete with a box hutch in the most private corner and a generous selection of perches giving good views in all directions (a couple of years later I would extend it by an extra yard or so). I met my new next-door neighbours, and warned them about my housemate. They seemed pleasantly intrigued rather than suspicious, and I looked forward to leaving the jurisdiction of Attila the Caretaker forever.

The notebooks remind me that it was on the morning of 21 August 1981 that the removal van drove off and I took a final look around the bare, dusty flat with Mumble on my shoulder. I had lived there for about fifteen years; it held plenty of good memories for me, and the big, high, sunny view through the window-wall was as alluring as ever. Well, time to go; I opened the basket for Mumble, and she obediently hopped in. On the way down in the lift I perversely hoped that we would run into the caretaker, so that I could smugly reveal who had been living with me for the past three years, but we didn't. We pulled out of the basement garage for the last time, and as we reached the open road across the old airfield Mumble made herself comfortable on the back of my seat, looking around her with interest.

During the drive I half-wondered whether, when I was actually living on the edge of a Sussex village, I might find that the reality did not live up to the expectation. Was I moving because I would really enjoy living in the countryside, or simply because I thought I ought to enjoy it? I needn't have worried. From that first night spent among the packing-cases I never missed my old flat for a single moment, and on all the evidence I can't believe that Mumble did either. For my part, I had achieved the Englishman's basic dream of owning his own tiny patch of one of the prettiest counties in the country, with a cozy pub just a few minutes' walk from my door. For Mumble's part, she had arrived on a new, green planet crowded with unsuspected fellow creatures.

8

◎◎

Mumble's Year

I T WAS HARDLY surprising that Mumble was disori-
ented by her new quarters for quite a few weeks after
our move to Sussex. I could only imagine that she must be
simply overwhelmed by the rush of new information bom-
barding her from every side. It was not just that her living
space was much larger; her surroundings were unimaginably different from her cage on the seventh-floor balcony.

For the first time—apart from her very brief visits to
Water Farm—she found herself at ground level amongst living greenery. In the aviary her hutch and the rear mesh wall
were tight against a hedgerow, from which I trained dense
tendrils of ivy inside to partly shroud her "private corner"
and nearest perch. But the hedge had grown out leggy, so
she had views in that direction through the green lacework,
across a field corner bordered by lines of hedges and trees.
Within a few yards in two other directions there were a big
oak and an old, ivy-covered plum tree, and beyond them the
rest of the late-summer garden and the back of the house.

Her floor was no longer of concrete carpeted with newspaper, but the living earth covered with grass, wildflowers,
and weeds growing in dappled sunlight and shadows. She
was no longer living in an urban cave: above her mesh ceiling, instead of a dark, enclosing slab of concrete she had the
open sky. What's more, it was a country sky—lively with
drifting clouds and flying birds by day, and by night black
and jewelled, instead of being stained a dirty orange-brown
by city lights.

"It was not just that her living space was much larger; her surroundings were unimaginably different from her cage on the seventh-floor balcony. For the first time . . . she found herself at ground level amongst living greenery."

Above all, on every surface of this whole new deep-focus environment, she was surrounded by constantly shifting patterns of miniature movements, while her ears picked up a vast range of sounds from near and far—the sounds of other living creatures, everything from insects up to cows. For a start, in the autumn and early winter the gardens, copses, and hedges around us were full of the rush and fuss of birds harvesting berries, seedheads, and nuts while they could, their numbers swollen by migrants from northern Europe. They attracted attention from other eyes than Mumble's; I never heard or saw a buzzard hanging above our spur of the South Downs, but one morning a hysterical chorus along the hedgerow across the nearest field heralded the big blue-grey bullet of a sparrowhawk exploding into the open amid a scatter of flying leaves.

For Mumble to assimilate all this from hour to hour, to process it and learn appropriate responses to it, must have been both engrossing and, at first, perhaps even scary (though I doubt it—Tawny Owls don't do fear). Throughout that September she was certainly hyperactive and distracted. She was difficult in the mornings when I let her out of the kitchen cage, and again when I took the basket out to the aviary to bring her in at night. She was also ravenous, but that may have been due simply to the season rather than any reaction to the move. I had noticed the previous September that for about three weeks she seemed much hungrier than usual, demanding chicks night and morning. Now she went into a chittering frenzy at the sight of food, and one morning she even dived straight out the opened door of the kitchen cage to snatch a chick right out of my hand as I approached. The weather was warm, but the only logical explanation I could think of was that despite the earliness of the season she was instinctively building up her fat reserves before the year turned cold.

◉◉

INSIDE THE HOUSE, I had set up her old night cage in a convenient corner of the kitchen, beside a window but round a corner from the main part of the room. She would be within close sight of the coffee-brewing facilities, so she would see a familiar activity when I uncovered the cage in the mornings.

I had taken the decision that with a spacious new aviary Mumble would spend more time out of doors than when we had lived in the flat, and that when indoors she would no longer have the run of the place, but would be confined to the big kitchen. I would miss our shared evenings, but there were simply too many crannies in the fairly open-plan house for her to hide (and crap) in, and—more to the point—too many windows that I would have had to remember to keep closed. Above all, I must admit that I didn't want my new home to be squalid with soiled newspapers and sheets of plastic.

The kitchen was large and, thanks to an extension, oddly U-shaped, with plenty of shelves and cupboards for her to perch on and explore, and its surfaces would be fairly easy to keep clean. I installed her tray-perch on a work surface at the bend of the U, from where she could see both halves of the room and both windows. Since keeping the door into the rest of the house closed would rob her of her usual vantage-point, I set up a perch for her at the highest point in the room, on top of a tall larder cupboard. This was close under the ceiling in the angle of two walls, much like her old door-top perch, with a good view to the largest window and the garden beyond. There were broad sills along both windows (complete with sacrificial potted plants), a double sink under the one facing the garden, and a large pine table that we could share. Nobody likes change, of course, but I was satisfied that all Mumble's main requirements for daily comfort were covered.

Since my commute to and from London was longer now, the chances were that sometimes I would not make the last train home, so I considered the possibility of designing some sort of owl-feeder that would deliver pre-thawed rations for a couple of days at a time. The sketches in my notebooks owe more to Rube Goldberg than to Leonardo, and, since I am no engineer, they seem to rely exclusively on melting ice as a timing system. (Measure time taken for ice-block of known dimensions to melt; place ice in plastic funnel of suitable calibre; place dead chick either directly on top of ice, or perhaps in small tray on pivoting arm counter-weighted at other end by ice-block; when ice melts, chick will drop—either directly onto the feeding shelf or, in one distinctly over-engineered version, onto an inclined plane that was apparently supposed to roll it somewhere else.)

Probably wisely, I went for a much simpler solution. I introduced Mumble to my new next-door neighbours on either side, and showed them where to find my spare key, ready-thawed chicks, and a small, framed feeding-hole in the mesh of the aviary above her dining shelf. After what must have been an initial shock, both Richard and Steve kindly agreed that one or other of them would step in whenever emergency threatened. They were as good as their word, and they never let us down—not even on dirty nights when it must have been quite obvious from my telephone call that I was in some London bar. I don't know what they thought of me, but in time they got quite fond of Mumble.

The diary records that it was 12 October 1981, five weeks

after the move, when Mumble seemed to have got a handle on the new situation and showed signs of reverting to her old habits. For the first time since we had moved in, when I opened the night cage that morning she did not play the drama queen, but hopped onto the doorstep perch, gave an alto croon, and put her face up for a good-morning nuzzle. Her appetite had stabilized, and she made no demand for breakfast. After spending some time on my shoulder while I made coffee, she flew up to her larder-top perch and settled down to rip the newspaper under it into tiny shreds, dropping them over the edge one by one and watching intently as they fluttered to the floor. We seemed to be back to normal.

◉◉

IT WAS ONLY when we were living in Sussex that I was able to make routine observations of the annual rhythm of Mumble's life. This was possible because I had achieved for her a vague approximation of a Tawny Owl's natural physical and mental environment—still very approximate, of course, but at least a good deal closer than had been possible in a flat high above city streets.

Over the course of the years that followed I noted both a definite sequence of seasonal mood-swings that coloured her behaviour, and the progress of the biggest physical event in her annual calendar—the moulting season each summer. As I have already mentioned, it was only after our move to the country that her moults seem to have conformed to an unmistakable and predictable pattern. The notebooks in which I recorded the daily changes and compared one year with the next inevitably make tediously repetitive reading, so the rest of this chapter is a collated compilation of the entries that I made over several years. It starts with the beginning of Mumble's year, a few weeks before a wild tawny would begin the process of reforging its bond with its mate and selecting a nest for that year's brood.

DIARY: I JANUARY

For the past three months she has behaved much as she was doing last October. This whole winter has been very mild, and she showed little sign of wanting to come in at night. I often left her out, and sometimes had to grab her when I did insist that she get into the basket.

She has started doing her midwinter "bat-walking" act across the mesh ceiling of the aviary. This involves flying up, doing a back-flip, grabbing the mesh with both feet, then walking "foot over foot" right across the ceiling while hanging almost upside-down with her wings slowly fanning. I cannot begin to imagine what it's all about, but there is definitely a rather aggressive swagger to the performance.

Today the weather is mild, dull, and rainy. I didn't get up until 9:00 a.m. (it's the New Year bank holiday, and most of the adult population of the British Isles are nursing a giant collective hangover), and when I came down Mumble was warbling softly into a corner of her kitchen cage. She kept this up until I uncovered the cage; then she turned and hopped to a perch, with a couple of soft squeaks. She waited quietly until I opened up, then jumped at once to her doorstep. She put her face up for a nuzzle, and kept soaking this treatment up for as long as I handed it out, rubbing and twisting her head against my face and gently pecking at my beard. She jumped to my shoulder when I patted it, and from there quietly to her tray-perch. She then sat calmly watching while I made my breakfast. She had a neat crap on the tray, then flew up to the larder-top perch for a bit of light self-grooming.

I made a few "Happy New Year" phone-calls, during which she squeaked obligingly when I held the handset up to her. This seemed to wake her up, and she jumped from my shoulder onto the kitchen table for a long, vigorous session of mutual preening. At one point we both watched a dog in the garden, with mild interest. She hovered close when, on an impulse (since she had crapped already), I opened the kitchen door and let her come through with me and upstairs, for a treat. This caused great excitement, but after spiralling up the stairwell she spent most of her time up there in the bedroom wardrobe, pursuing her obsessive interest in dark burrows.

When I eventually took her downstairs she hopped obediently into the basket. Once inside the aviary she went into her hutch briefly, then came out onto her doorstep for ten minutes or so while she checked out her domain. When I looked out the window at about 11:30 a.m. she was inside the hutch, and she seemed to stay there silently for the rest of the day. I heard a few tentative hoots at about 9:00 p.m. I brought her in at 11:30 p.m., without any objection; and so to bed.

2 JANUARY

This morning there was no "necking." She jumped straight to my shoulder, then to the tray-perch, then up to the larder-top; and there she stayed, resisting all invitations and blandishments. As it was a vile day, with blowing rain, I left her there when I went out shopping. When I got back at about lunchtime there was evidence that while I was gone she had come down to the tray-perch, used it, and returned to the cupboard (if only this was a predictable daily routine . . .). Again, there was no difficulty persuading her into the basket, and once in the aviary she went straight to her "private corner" perch and stayed there.

For the first time this winter there are sheep in the nearest field; she seems completely uninterested in them.

5 JANUARY

When I put her in the kitchen cage last thing, it became clear that she had secretly stowed a bit of her last night's chick under the newspaper in one corner for later—this is the first time she's done that this year. I still gave her a whole chick tonight, and she finished both.

SECOND AND THIRD WEEKS OF JANUARY

She's still hiding a breakfast snack from her supper about every second or third night. This has roughly coincided with a change to much colder and windier weather, but there seems to be no exact correlation.

More noticeably, she has begun doing her "hoot and head-shot" routine, as in previous winters. Is this change of behaviour connected with the mating season? When she first hears me morning and evening, there's a great deal of "Indian whooping into corners" inside the night cage and aviary hutch before she emerges into sight. Then, when I let her out in the morning, when I go into the aviary at night—and on a few occasions when I have come back into the kitchen unexpectedly while she was free—she hoots quite an aggressive challenge and flies at my head. She doesn't strike at it with her feet, just lands on top. When I put up a wrist she steps onto it and lets herself be carried down quite calmly, and there's no repeat of this behaviour while she's free. (So far, there's no sign of the next recorded stage of the midwinter behaviour— the "whistling war-dance.") She's still generally quite sociable; she came to my shoulder unbidden during a Sunday-morning newspaper session, then jumped to my crooked-up knee and craned up her face for a nuzzle in the old way, talking in soft squeaks and croons.

FOURTH WEEK OF JANUARY

She's now doing the full-blown "HHS+WWD" when we meet morning and night. First the warbling into corners, then the "hoot and head-shot," then the "whistling war-dance"—as I lift her down from my head she climbs up my arm to the crook of the elbow and kicks at it a couple of times, while giving whistling squeals and flapping her wings in an excited little spasm. And whenever I go into the aviary at night she does a lot of vigorous "bat-walking" across the ceiling, ending with the most amazing back-flips down again, to land sitting neatly upright on pinpoint targets like the hutch doorstep or the inside of the basket. Given the angles and distances, these three-dimensional aerobatics should be completely impossible; she makes Russian teenage gymnasts look clumsy.

MUMBLE'S "BAT-WALKING" ON her ceiling was standard behaviour each midwinter, but the onset of the hooting and head-shot plus the whistling war-dance differed slightly each year, the first element always preceding the second by a week or more. I noted them in mid and late January respectively in most years, but in 1988 she remained positively dozey and cuddly until the last week of January, then began the HHS tentatively and inconsistently, and the WWD didn't kick in until the first week of February. (Once she started these behaviours in any year, however, she kept up the full routine until about late May.)

At about this same time each year I also noted that she was caching part of her supper overnight, and bringing it out of the night cage with her in the morning. She then usually carried it around the kitchen in her beak from perch to perch, bugling insistently, until she finally settled to eat it on top of the larder. I always offered her a feed night and morning during cold winter weather, and she never turned down the chance of a chick. But while I didn't make the connection at the time, I guessed later that the snack-saving might be because the twice-a-day feeds weren't allowing her enough time between meals to digest a whole chick. I found it impressive that she realized this in advance, and had the self-discipline and foresight to stop eating half-way through and tuck the rest away for later.

In 1989 I recorded that the timetable was slightly delayed compared with the previous year. On 28 January I noted that she was doing the HHS at night but not in the morning, and there was no sign of WWD. She still liked her nuzzle on the doorstep first thing, but would only come to my shoulder or lap and demand further preening after she had been free in the kitchen for at least an hour during long weekend mornings. (The notebook recalls that one morning she was sitting on my shoulder while I was reading the paper when she suddenly did a titanic sneeze, spattering the newspaper two feet in front of her "nose." She shook her head vigorously a couple of times, but then—like me—she calmly turned her attention back to the *Sunday Telegraph*'s thoughtful analysis of the Five Nations rugby tournament.)

That January she showed no desire for breakfasts, and late in the month she had not yet started caching snacks overnight. I wondered whether the mild winter might be affecting her behaviour; we had had very few frosts that year, the crocuses were up, and even a few foolhardy daffodils. However, in mid February 1989 I noted that she was again stowing away snacks every second or third night, and by the third week of the month it was every night—though this seemed to have nothing to do with how well I was feeding her. When she brought the snacks out of her night cage with her, she seemed to be trying to tell me something. She kept following me around with them, even bringing them to my shoulder, while giving the full hoot—"*Hooo!* . . . hoo, hoo-hoo *HOOOO!*" (With her mouth full, though, this sounded more like quacking.) She seemed to expect me to *do* something with them. Was it possible that she was trying to feed me again?

In 1991 I noted that she didn't start the hooting and head-shots until the second week of February, and there was no whistling war-dance on my arm until the middle of that month. Again, I've no idea if the weather was a factor that year; we had an unusually severe winter, with heavy snow and sub-zero temperatures for a week in early February. Generally, this didn't seem to inconvenience Mumble at all; she was not really bothered if she came in for the night or not, and as soon as the water in her dish melted she had a thorough bath. I found her sitting in an icy wind, soaking wet right through but apparently quite comfortable.

THIRD WEEK OF FEBRUARY

The pattern seems firmly established: "bat-walking and back-flips," saving snacks every night, and "HHS" plus "WWD." Her emotions are at a high pitch, and her feelings seem confused (I'm tempted to compare this with PMS . . .). One night when I brought her in she jumped out of the basket and flew straight into the night cage, but when I approached with a chick she flew straight out again and hovered round me at waist height, wings flapping, until I threw it in.

5 MARCH

Much the same morning drama. She emerged carrying a full half of last night's supper, and flew around from perch to perch, keeping close to me and all the time yelling monotonously round her beakful of chick. This went on for nearly ten minutes before she finally took it up to the larder-top and ate it. I wish I could figure out what she expects me to do about this—I feel stupid in the face of her urgent but incomprehensible nagging. Does she want me to take it from her? Is this some sort of displaced maternal behaviour? At this time of year it's impossible to interpret what role she has cast me in, and it seems to change from one moment to the next.

When she finally finished her snack she seemed to calm down. She stropped her beak clean a few times on the "cliff edge," then settled down—but not on her perch. She came right forwards to the edge of the larder cupboard and lay down flat on her front, with her breast and shawl feathers puffed up and her face resting on top of her claws. But not more than thirty seconds passed before she looked at me, did a theatrical double-take, and her eyes grew huge. She started hooting aggressively, shuffling from one foot to the other, and then flew at my head. When I put up an arm and diverted her down, she didn't do her "whistling war-dance" in the crook of my elbow, but clung to it tensely, mantling her wings and giving occasional hoots. She let me kiss and cuddle her for a moment, then flew off high again—and soon afterwards, back into the cage, where she started her monotonous whooping into a corner. This has *got* to be something to do with the nesting season?

LAST WEEK OF MARCH

The signs of confused restlessness continue, and she's still stowing away a snack almost every night. Very frequent "hooting and head-shots" followed by "whistling war-dances," but I've noticed a refinement of this. Instead of just kicking my arm, she pumps her body against it, violently and repeatedly—surely this must be sexual? (Though it seems more like male than female behaviour.) There is a lot of noise and ricocheting around in the aviary if she happens to see anyone or anything moving, even on sunny afternoons when she should be sleeping.

I notice that I have a robin's nest in the hedge; as they go about their endless hunt for food she pays no attention to them, nor they to her.

11 APRIL

Same noisy, rough behaviour, but this morning, for the first time in weeks, there was no sign of a cached snack. At night she headed into the cage like an arrow as soon as I opened the basket, and into a corner to start her "Indian whooping." When I threw her supper in she scuttled across to grab it—whooping on the approach, and quacking brassily on the retreat with it in her beak—and after she'd eaten it I heard her singing out challenging hoots for quite a long time after I'd gone to bed. I've noticed a couple of small down feathers on the night cage floor—too early in the year to be significant? Still, at the same time as all this drama she will tolerate the occasional, brief nuzzling session, so long as I don't push my luck.

7 MAY

Over the past month there have been steadily fewer snacks cached overnight, and it's now down to about one every third day. There's still a lot of noise, day and night, and a lot of "HHS" and "WWD" when we meet. But if I refuse to take all the dramatics seriously she can sometimes be talked out of it by my nuzzling her head during the "war-dance" phase, though she's pretty grudging about it, and seems distracted.

10 MAY

Only one snack cached over the past week. The usual "HHS" and "WWD" when I let her out this morning; but—for the first time since the end of January— this was followed, after a thoughtful pause, by her instigating a proper, fifteen-minute, old-fashioned mutual preening session at the kitchen table. She repeatedly came, unbidden, to my shoulder, lap, or the table under my face, and there was much pleasurable head-barging and nibbling, enlivened at intervals by pounces into my newspaper.

22 MAY

This morning she did the usual "whooping into a corner" when I came into the kitchen; but when I opened the cage she came to her doorstep and sat quietly, blinking and craning her face up for the first proper "good morning" caresses for about three months. The night routine wasn't quite so aggressive, either. When I went into the aviary she did the "HHS" and "WWD" routine, but only briefly and half-heartedly, as if simply because she felt it was expected of her. Then, instead of "bat-walking," she jumped promptly into the basket.

◎◎

THE DIARIES SHOW that this change of behaviour from rough, fidgety, and distracted to positively friendly began between 22 and 27 May every year. Once it began, it was consistent: she was calm and sweet-natured, expecting a nuzzling session first thing in the morning, and when we spent hours together at weekends she approached me often, demanding more. She hardly ever cached bits of her supper overnight, and never after 1 June; instead, she demanded a whole chick each breakfast time as well as one most evenings (naturally, I had been dunking her rations in supplement powder for a couple of weeks already). Although the hooting and head-shots and the whistling war-dances might recur very intermittently, they were increasingly brief and half-hearted, and I never noted them later than 31 May. One year I recorded that on 24 May I saw her adopting her "broody egg-sitting" pose on one of the kitchen work-tops, but this was brief and unrepeated.

By these last ten days of May I was watching out for her to begin her moult, but I noted very little feather loss, and then only small contour feathers. The follicle of a living feather has a blood supply, and when the old feather falls out a new one starts to grow immediately from the same follicle. The replacements emerge as "pin feathers," which are tightly furled inside a thin shaft that soon splits open to allow them to grow and deploy. The earliest Mumble ever lost a flight feather was 26 May one year, when I found a secondary in the night cage; another year she lost one on the 30th, and a pair of primaries and the matching secondary during the first week in June.

I knew from the textbooks that Tawny Owls have this impressively clever pattern of moulting. The loss and replacement of their flight feathers begins at the inside ends of the rows of primaries and secondaries, and they lose the same feather from each wing within a few days, so that their flying balance isn't upset for long. In the wild, how many they lose that year depends on how well they are feeding, but in any case the moulting process stops in September. The following spring it resumes, at exactly the point among the sequences of feathers that it had reached when it stopped the previous year.

Mumble often lost her first flight feather during the first three days of June—once, in a matched pair over the same night. Occasionally this didn't happen until the 10th or 15th, and (uniquely) in the very hot summer of 1986 I didn't note it until 25 June. Most years the diaries record that she was dropping feathers like snow by 20 June—two or four wing and tail feathers and half a dozen body feathers every twenty-four hours.

◎◎

WITH THE BEGINNING of the moult proper—and precisely coinciding with it, whether it began early or late—Mumble's character showed a profound change. She became a different owl, not only physically but emotionally. She made very little noise in the aviary during the light summer evenings (perhaps because her condition caused a loss of confidence?), but once I brought her in she would spend a lot of time denned up in the open night cage, whooping quietly to herself. It was most noticeable that from being merely calm when she was with me, her mood turned positively clingy. On the first day of June one year I noted: "Very subdued, insecure, and *very* cuddly—just like a child that feels unwell. She couldn't keep her beak off me for five minutes at a time; she demanded nuzzling first thing in the morning, then every few minutes for well over an hour. Lots of quiet crooning when she comes over to me."

Every year, this was the consistent pattern throughout the three months of the moulting season. I was long past the cynical suspicion that her approaches simply meant that she was itching and wanted a scratch; she was perfectly capable of scratching and nibbling herself all over, with far defter and sharper instruments than my nose. I believe it is plausible that she felt a need for comforting reassurance during the long weeks when she could feel that she was in much less than peak condition, and there is some scientific support for the theory that mutual preening in birds reduces stress hormones.

(However, honesty forces me to note that in the wildlife studies that I have read I have found no mention of this behaviour between pairs of tawnies in the wild. June is early

Just a few examples of Mumble's huge variety of plumage. It ranges from the tiny vertical feathers from the crown of her head (top left) to variations of the downy feathers that covered the front and underside of her body. (Photograph: Edward Reeves/Lewes)

for a wild tawny to finish caring for its fledgelings, and common sense suggests that it would be impossible for moulting to start while the parents were still engaged in the exhausting hunting cycle necessary to feed them after they begin exploring outside the nest. When the owlets are finally driven out of the parents' territory later in the summer the parents always separate for a few months, rather than "being there for each other" during the moult.)

The need to keep up her energy to build new feathers was presumably behind the fact that on 3 June one year I noted in the evening that she had caught a small mouse in the aviary during the day, and on 6 June her "game book" records another one. On the evening of Wednesday, 7 June, I found her carrying the headless remains of a sparrow around the aviary—it was fully feathered, and must have flown in through her feeding hole (a slightly macabre thought—little did it know . . .). On Saturday, 10 June—after noting that she had lost a couple of covert feathers, had been frantic for her breakfast chick, and was still demandingly affectionate at every opportunity—I recorded: "Either she kept her Wednesday sparrow cached in her hutch this long—which seems unlikely, given her sharp appetite at the moment—or she caught another for her tea. Usual muffled squawking with her mouth full sent me out at about 6:00 p.m., and there she was, carrying it around on the usual circuit of honour." (Mumble's hunting behaviour is described in the next chapter.)

◎◎

17 JUNE

In the past week she has lost another couple of second-aries. Her mood remains needy and affectionate, and she's at her most charming. She has also been a bit more energetic and adventurous. This morning she played games among the legs of the chairs pushed up to the kitchen table, hopping precisely along from one of the stretchers to the next with fast horizontal jumps—there's no room down there for her to spread her wings even half-way. After she got bored with this and flew up high again, she suddenly landed on the newspaper I was reading on my lap, and sat there demanding a nuzzle. When this was over she sat quietly for a moment—then POUNCED into the middle of the newspaper from a standing start. She proceeded to attack it with loud, almost alarming force, wings half open, dancing on extended legs and then SMASH-ING down with her feet. When I complained, she flew up to the larder perch and carried on her mock battle, her claws now ripping like iron hooks at the "cliff-edge" of the plywood top of the cupboard.

28 JUNE

Moulting is fast and furious now, and she is consis-tently keen on a substantial daily breakfast in addition to her night-time chick. I left her out for the past two nights, and this morning when I went into the aviary I found two perfectly matching primaries—left and right mirror images—lying about a foot apart; the "selective symmetrical moult" process is unmistakable. She is taking extra baths, but since the weather is very hot I don't know if this is specifically to do with the moult or just common sense.

7 JULY

For the past week she has been losing one or two primaries and secondaries and many smaller feathers every day. The last tail feather went last night—she has only her big cone of downy white "bum-fluff" left back there. Her landings have become noticeably clumsier without her rear air-brakes, just like when she was first learning to fly in the flat. The mood of clinging insecurity and frequent demands for affectionate preening are fairly constant while we are together.

11 JULY

Among the downy stuff on her bum, three or four filmy feathers seem to be getting a little longer and showing a pale brown central stripe. She is very touchy about them, and chittered and moved when I gently stroked them. [I found out only later that a bird's new pin feathers are very sensitive.] She is still bathing more often than usual, twice in the last three days— once in the aviary dish, and once in the kitchen washing-up bowl—and always leaves a few small feathers in the bathwater. Afterwards, her method of locomotion through the air cannot really be called flying. By an immense expenditure of effort she blunders vaguely from perch to perch, making a noise like a wet flag in a high wind, before arriving with all the elegance of a flung dish-cloth.

14–16 JULY

She's lost her left wing main primary—the longest feather on that wing; the equivalent on the right is still hanging in there. There's a big cone of new, pristine white fluff on her bum, and at the top of it about nine potential new tail feathers now seem to be identifiable in a spade-shaped clump. The tips seem to be getting less filmy, and the brown brush-stroke down the middle is getting more prominent, but they show no real sign yet of developing a proper quill structure.

22 JULY

This morning the overnight cage had half a dozen medium-sized body-feathers, a dozen of her tiniest face feathers, and a neat drift of "quill dust" that nearly filled a teaspoon. She loses one or two primaries and/or secondaries each day, and a mass of smaller feathers. She seems to be quite hungry, but not frantically ravenous, and I continue to dust about every second chick with SA 37. Her mood is still very cuddly and needy; she jumps to me at every opportunity, with a lot of quiet crooning and squeaking.

27 JULY

She has now regrown a lovely, symmetrical fan of proper, quilled tail feathers above the cone of down.

They still seem fractionally short, but she can fly well in all configurations.

Throughout July, while I've mostly been watching this going on at the stern, the steady loss and replacement of other feathers has continued. She does a lot of scratching, and looks pretty scruffy, with half-shed feathers sticking out of her body and wing surfaces at odd angles for a day or so until they drop or she pulls them out. She is still bathing a little more often than usual, and is still dependent and demandingly affectionate.

8 AUGUST

For the past couple of weeks she's been very hungry— she yells for breakfast every morning, so I usually give in; she knows best, after all. Her new tail is now full sized and perfect in all respects. She is still occasionally dropping a big wing feather, and she is shedding back feathers too. Her head looks very scruffy and spiky, and this seems to be the main scene of action at the moment. It gives her a "flat-top" look, like a military buzz-cut from the top edge of her facial disc right back to the top of her shawl feathers—this makes a weird, belligerent-looking contrast with her generally fat, fluffy appearance.

17 AUGUST

We've had a very hot spell recently, in the 90s F; it has now broken, but it's still in the 70s in the middle of the day. Despite this, when I carried her out to the aviary this morning I caught the first whiff of autumn on the misty air (extraordinary how unmistakable this is, even to a smoker like me). There is still some moulting and replacement going on, but the main flight feathers seem almost complete. Mumble seems less hungry; she's rarely interested at breakfast time, so I've cut out the morning chick. Today—after an unbroken month of her positively demanding a necking session on her cage doorstep in the mornings— she "whooped in the corner" when she heard me come into the room, and when I opened her door she did not immediately bounce out. She's still affectionate, but takes her time before making advances to me.

22 AUGUST

Her head feathers are now fully replaced, and the main action over the past couple of days has definitely been on her front. There are blizzards of fluffy breast feathers in the overnight cage, and after she's had a bath or there has been a rain shower they stick to the mesh of the aviary walls, fluttering like little prayer-flags or Mongol standards. The faint signs of a mood change are now definite; she still accepts mutual grooming sessions, but she no longer instigates them herself.

7 SEPTEMBER

Apart from everyday scratchings the moult seems to be over. Her mood is still friendly, but more independent. I'm going on holiday for the next couple of weeks, and I fully expect that her behaviour will have changed completely to "independent autumn mode" by the time I get home.

20 SEPTEMBER

Got back from ten days in Switzerland and France. [Thanks to my old friend Gerry, this had involved many opportunities to fire a replica fifteenth-century cannon, and a lot of uproarious drinking, singing, and laughing among like-minded spirits crowded around enormous camp-fires.] During my absence I had arranged for three- to four-day ration packs to be delivered to Mumble in the aviary. She's in perfect feather all over, and looks wonderful. She's friendly in a casual way, but not really interested in being handled. Her morning greetings are perfunctory, and she prefers to go off and sit by herself, holding inner conversations of soft, querulous warbles.

Last night there was a good deal of calling back and forth with a wild owl—she did the classic " kee-*wikk*s!" in answer to "*Hooo!* . . . hoo, hoo-hoo *HOOOO*s" out in the trees a couple of hundred yards away.

27 SEPTEMBER

Over the past few days she has been becoming more nervy and active. She "whoops" into the corner of her night cage when she hears me come down each morning, and while she jumps out onto my shoulder she won't stay there for long. She shows no sign of wanting closer contact on most days, and only wanders over for a brief nuzzle on weekend mornings after she has been free in the kitchen for a good hour or so. Her appetite seems to have sharpened up again, night and morning, and she zips smartly into position at my first "come-and-get-it" whistle.

◎◎

THE NOTEBOOKS RECORD that one particular autumn her toughly independent mood then went into reverse, for no reason that I could understand. For about three weeks between late September and mid October 1990 she reverted to quiet, gentle, hesitant behaviour. She allowed demonstrations of affection (within reason) first thing in the mornings, and would come to my patted shoulder morning and night for brief sessions of mutual preening before flying off. I noted that while paying attention to her surroundings alertly, and flying smartly from the basket to the night cage as soon as I brought her indoors, she showed a reduced appetite. Even so, while she never asked for a breakfast chick, equally she never cached an overnight snack.

Often she showed no interest in coming in for the night, so I would feed her in the aviary and leave her there. One wet, chilly night in the third week of October I went out to her at about 11:30 p.m. during a gap in the showers. When I saw her shaking herself damply on her private perch, I thought she would be eager to come to the basket. Instead, she looked at me with a complete lack of interest—then jumped down to, and immediately into, her water dish. After doing the full ducking-and-flapping routine, she climbed up to sit on the rim while she shook herself a bit—and then jumped, splashily, straight back into the water. This demonstration of her complete independence, whatever the weather, was slightly crushing: dismissed, I humbly laid her chick on the shelf and left her domain.

By November she was back into her unvarying winter routine, which is best described as "stereotypically British"— she was perfectly civil but fairly self-contained, and only occasionally affectionate. Each morning she would warble softly into the bottom corner of the night cage when I uncovered it. She would come hesitantly to the doorstep, where she seemed to enjoy a brief greeting, but then either went back inside to her corner perch or hopped to my shoulder and then flew directly up to her high larder-top. Usually she stayed up there for as long as I gave her the chance, but most weekends she would occasionally fly down to my shoulder or the kitchen table after an hour or so, diffidently suggesting a necking session.

When it was time to get into the basket for the trip to the aviary she was quite calm. She spent most of her winter days in a motionless ball of feathers in the dark corner of her "private" perch, camouflaged by her ceiling and two walls of ivy. In the evenings she came to my shoulder briefly, but then usually had to be coaxed or even grabbed to make her enter the basket. Her mood was generally low-key, but perfectly pleasant. There was no bat-walking, no whooping and head-shots, and no whistling war-dance. For the great majority of the winters we lived our lives peacefully in parallel.

◎◎

IT HAD BEEN predictable that our relationship would become rather more distant in our new surroundings, not only literally but also psychologically. At its simplest, the reason was that Mumble and I were no longer spending almost every evening in each other's close company.

After our first year or two of excited mutual discovery in the flat, our relationship had anyway settled into a contented routine, but while we were sharing the same space most evenings I had still been the main focus of her attention, as she was of mine. But in Sussex we no longer really "shared a territory" except for during weekend mornings; she had her own territory—and a new, crowded mental life—out of doors. It was quite natural that her much more varied surroundings and the stimuli that they constantly delivered would occupy her attention for much of the time. For me, the great compensation was the opportunities they gave me to watch her adjusting to these new circumstances and excitements.

9

◎◎

Real Trees and
Free-Range Mice

I HAD A certain amount of adjustment to cope with myself, and that first winter of 1981 brought one temptation that I had never known in the city. There was a large private estate nearby, and the fenced wood on a ridge above my village was full of reared pheasants. On several mornings a rusty, clucking call outside my window betrayed a handsome cock bird strutting around in my garden as if it owned the place.

On leaving London I had given up my membership in a pistol club and my licence to keep full-bore firearms. It had been something of a wrench to part with them, particularly my antique 1896 "broom-handle" Mauser. (It was horribly over-complicated to strip and clean, and uncomfortable in the hand; but it was delightful to fire with the wooden holster-stock attached, and it oozed historical charisma— this was the type the young Winston Churchill had carried in the cavalry charge at Omdurman.) It just wasn't practical to keep up with this particular hobby; I had enough to do settling into new routines without the trouble of finding another club within a reasonable distance, and of fulfilling the legal requirements (which were onerous, even before private pistol ownership became illegal in Britain). However, the move did make it easily practical for me to legally acquire a shotgun, and I soon began to enjoy the new challenge of clay-pigeon shooting.

The gun naturally had to be kept in a locked steel cabinet

bolted to a wall, and although my trigger-finger itched whenever I saw pheasants in my garden, I felt obliged to ignore the impulse. They were my lawful prey—my land, so my bird, and just beyond the legal limit of fifty yards from a public road—but it was naturally out of the question to sub-ject Mumble's ears to the shock of gunfire at close quarters. And anyway, the potential dinner would surely be gone long before I could unlock the cabinet, get the 12-bore out, and sneak around the side of the house. The pheasants didn't always get a free pass, however. On a later occasion one of these loud fools made the mistake of lingering in the morning when Mumble happened to be safely denned up in a wardrobe upstairs. Even when I shouted, the intruder's take-off lacked urgency; a couple of weeks later, roasted under a wrap of fat bacon, he was delicious. (The lady next door had been a bit startled by the shot, but as I wiped up bread sauce with the last forkful I felt a primitive satisfac-tion over exercising my territorial rights.)

In the wild, winter has both advantages and disadvan-tages for an owl exercising its rights over a much wider territory than my few square yards of garden. In the biting cold a tawny needs to keep up its strength with very regular feeding. It helps that there is less food about for the rodents; they grow careless during their constant foraging, and the thin, pale grass gives less cover for the runs of the field voles. On the other hand, if snow falls and lies it hides their tunnels beneath the surface. Although an owl's hearing is sufficiently acute for it to stalk rodents successfully even when they are surprisingly deep below a layer of snow, it is in weather like this that small birds living within an owl's territory are wise to pick their concealed night roosting-places with particular care.

DIARY: 20 DECEMBER 1981

For the past two days and nights it's been snowing quite heavily off and on. Although Mumble's "private corner" perch is well sheltered from above, the snow has been falling through the mesh roof thickly enough to make a drift several inches deep on the feeding shelf, and the water in her dish is frozen solid. She'll have to get used to this every few winters, so I take her out as usual in the mornings. Snow and ice is a first for her, and she seems unimpressed. She tried standing on the ice in her dish, and pecked at it a couple of times before giving up in disgust. Then she paddled around in the snow on her shelf, and tried to eat a mouthful of it. This was definitely a mistake, and she shook her head furiously to get rid of it—*uggh!* She spends the days as normal, fluffed up inside her built-in duvet on her private perch, and doesn't seem to feel the cold unusually (after all, she's built for it). Her appetite is healthy, but not ravenous. I bring her in early in the evenings; last night I tried leaving the kitchen tap dripping in case she was thirsty, but she preferred to drink from a saucepan that I had left full of water to soak clean in the sink.

9 JANUARY 1982

Saturday afternoon—I was working in the office [the back bedroom overlooking the garden] when I heard a lot of unusual noise, including the "belling" of hounds and the thudding of hooves. When I leaned over to look out the window I could see flashes of bright scarlet beyond the nearest hedge. I got outside in time to see the local hunt crossing the fields at the bottom of the garden, and a stream of noisily excited foxhounds poured through a hedge-gap within a couple of yards of the aviary. They were not visibly chasing anything (I've never seen a living fox around here), but the soundtrack was splendidly stirring.

Mumble reacted by getting onto a perch as far away from the action as she could. I've read that birds are more disturbed by vibrations through the ground than by actual noise—could the thunder of the hooves have been transmitted up the embedded branches that support her perch? Then a riderless horse ambled up and put its head over the fence, and Mumble went into "pigeon-spotting mode"—all tall, thin, and suspicious. The horse was eventually followed by a stout, puffing gentleman dressed in mud-caked black and white, who took some time to gentle it before leading it away. By this time the hounds were running in happy, pointless circles all over the field. Mumble didn't return to her usual perch until the whole noisy pageant had drifted away. When the hounds were just a quiet song on the breeze she eventually settled herself flouncily, like a scandalized maiden lady.

12 JANUARY

Cold, starlit night. When I went out to collect Mumble I couldn't see her until I was inside the aviary—she was sitting absolutely rigid and focused at one end of her private corner perch. Eventually I heard the very faintest of rustling noises in the grass at the foot of the hedge. Calls, whistles, stroking, and even blowing up her breast-feathers (usually guaranteed to provoke her) failed to get the slightest response. Since she was facing away from me, when I got cold and irritated enough I just picked her up from behind her legs and put her in the basket. When we got into the kitchen she flew to the top of her night cage and glared out the side window, staying quite still and visibly on guard. Eventually the offer of a chick tempted her to forget the potential meal and enjoy an actual one.

16 JANUARY

We enjoyed the usual leisurely Saturday-morning routine in the kitchen. While I made my breakfast she was having hers in the open night cage. When I heard her thump up onto the doorstep perch I went round, and she greeted me affectionately—perhaps a shade too affectionately, since she hadn't yet had a "feek" to clean the goop off her beak. Then she jumped onto my shoulder and came back to where I was sitting at the table with my second coffee. Just as when we were in the flat, this is our main time together, and we had an enjoyably vigorous snogging session.

This always puts me in a fond mood, and in a moment of indulgence (after she had reassured me by using her tray-perch copiously) I took her upstairs with me when I went up again to shower. She took off and flew up the high stairwell in a tight spiral, meeting me

Mumble enjoying exploring a bookshelf during one of her rare visits to my office. The more cramped the conditions, the better she liked it, sometimes even burrowing behind the books and shoving her way along the back of the shelf.

on the landing. Although I left the bathroom door open for her, she was much more interested in playing on the bookshelves on the landing and in my bedroom. The only price I paid for letting her upstairs was that she crapped over a Filofax that I had carelessly left lying around (a mishap that I saw as a powerful image of the conflicting demands of my urban and rural lives).

I cannot fathom her weird passion for crawling head-first down tunnels, apparently the narrower the

better. She crouched flat on her front to squeeze onto
the top, half-depth shelf on the landing (five-inch clearance), and at one point she was actually crawling along
behind the books on another shelf, warbling happily.

◉◉

BY THE SPRING of 1982 Mumble seemed to have entirely
settled into country living. Although her normal calm was
disturbed whenever she saw something new for the first
time, she soon became blasé about it and simply filed it away
in her head. An exception was the sound (or vibration?) of
farm machinery close by. She seemed to dislike ploughing
and harrowing in the next field, and particularly the racket
of a tractor with a hedge-trimmer attachment as it thrummed,
crashed, and screeched its way along the hedgerows.

A good deal of this learning process must have gone on
while I was out at work, but occasionally I was on hand to
notice her doing her romaine lettuce imitation. The first
time a Friesian cow put its head over the fence a few yards
from the aviary she seemed to go into a rigor of stunned dis-
belief, squeezed tight and thin and stretched to about half
again her normal height, with a furry hatchet face and Chinese eyes. One May night when I went out to fetch her I
heard a fearsome wheezing and grunting in the garden, and
found Mumble rapt and motionless on her front perch. I recognized the noise from a memorable night during my sub-
urban childhood, but she didn't; she was utterly transfixed
by the spectacle of two hedgehogs copulating laboriously
under the raspberry bushes.

I expected that in the mating season she might get rather

more gentleman visitors than she had done in South London; after all, there were more of them around down here (and she was easier to find). As it turned out, the frequency of such advances seemed much the same—and so did her responses. During February and March her mood was generally belligerent; she was permanently on the lookout for intruders, wearing her war face. Her sense of her own territory must have been much greater in the Sussex countryside than it ever had been on an urban balcony. She had much wider views, but she had to compensate by calls and displays of body-language for her inability to fly out and physically defend this wider "mental territory."

I heard plenty of calling and counter-calling on winter nights, but I never happened to see a male tawny at such close quarters as I had done when in the flat (though I might well have simply missed such encounters—obviously, I was much further away from the action when Mumble was in the aviary at the bottom of the garden). However, I did make a note of one rather more surprising episode.

When I went out to fetch Mumble one night she was entirely calm and friendly, until she suddenly jerked alert and stared vertically up through her ceiling. She leapt out onto the most exposed of her perches and kept up this intent surveillance, twisting, raising, and lowering her head as her computer locked on to whatever she could see and hear up there, and tracked its every move. Clearly, something that I couldn't make out was circling above us in the dark. I waited for her to bellow a challenging hoot or a vulgar "kee-*wikk*!" but to my surprise she remained silent. Then, from the darkness above there came an eerie shriek— the intruder was a Barn Owl, not a tawny. Still Mumble stayed silent, but her head tracked it smoothly as it flew away. It took me a long time to settle her down before she would get in the basket and come indoors.

◎◎

DIARY: 3 JULY 1982

I simply never can tell what will excite Mumble, and what will leave her unmoved. On Sunday afternoon I was sitting reading in the garden when I saw Buster [a neighbour's cat] picking his way down the back lawn towards the hedge-gap leading into the field. Mumble came out of her private corner onto the front perch, and fixed her attention on him. Cat and owl watched one another respectfully and at length, like a pair of wary gunfighters in a Sergio Leone film, but Mumble showed no actual excitement. Her mood was one of calm, still, focused watchfulness: "You and I both know that we don't want to tangle—so just keep on going, Buster, and we'll each tend to our own business . . ."

On the other hand, one afternoon recently when I walked past the aviary carrying a bunch of sweet-peas that I had picked, she stirred from her doze into a display of apparent outrage. This happened again, several times—and always with sweet-peas (go figure . . .).

◎◎

OCCASIONALLY, BUT NOT often, Mumble would be "mobbed" by small birds who spotted her in daytime, despite the ivy camouflaging her in her dark private corner. From my desk upstairs I would hear this begin, usually with the repeated alarm call of a blackbird, and by the time I had got to the window to look down, a number of smaller birds— sparrows, finches, and tits—would be gathering as near to the aviary as they dared. They moved around nervously from perch to perch, flapping their wings and cocking their tails, while delivering a cacophony of alarm calls to adver- tise Mumble's presence to any other potential prey. This raucous scolding might go on for several minutes at a time, but she seemed able to ignore it completely—she might blink, but she didn't even shuffle on her perch, and certainly didn't emerge from her cover in a threatening way. Appar- ently this is normal even in the wild, where there is no wire mesh between the owl and its tormentors. The racket has only a nuisance value, but I was impressed by Mumble's patience—I got more irritated by the monotonous hysteria than she did.

◉◉

I WAS VAGUELY disappointed that the first actual kill I saw her make was an earthworm. I had read that tawnies often hunt worms on the ground when damp weather brings them to the surface, but although I congratulated her when I disturbed her eating it one dreary day in March 1982, I couldn't help feeling that this was nothing to write home about. I supposed that it was as much as I could real- istically expect; a worm on the surface would be easy to catch, and was not entirely unlike the shoelaces she had loved playing with as a fledgeling. More significant, Mumble had never been given the slightest education in the subject of brown furry things that ran about on four legs; if one scuttled across the aviary how would she even recognize it as supper, let alone seize the brief opportunity to catch it? (Incidentally, worm-loving readers may be consoled to learn that throughout the speedy process of ingestion the luckless creature maintained an air of stupefied boredom, displaying no more animation than the length of spaghetti that its final exit so closely called to mind.)

However, within the month Mumble proved that she had indeed taught herself how to kill for food without the aid of instruction, and that a fairly steady stream of the local rodent population were carelessly taking shortcuts through the aviary. In good weather I quite often left Mumble out at night if she seemed reluctant to come indoors, so these fatal encounters may have taken place more regularly after dark, leaving no evidence at the scene by the time I came out to her the next morning. Whether or not that was true (and it seems most likely), the kills that I was aware of seemed mostly to take place during the day, and usually during the spring and early summer. This pattern repeated itself every year from 1982 onwards, and the following notes are a compiled selection from her "game book":

MID MARCH 1982

First-time kill? I went out to the aviary late at night, and as soon as I opened the back door of the house I heard her giving muffled, brassy whoops, as if she was playing a trumpet muted with a rolled-up pair of socks. She had her mouth full, of course—with about 75 per cent of a self-caught mouse. She was very proud and fierce, but seemed a bit undecided about what to do next—after all, if this *is* her first time, she has never "been on the course." She brought it with her into the basket, and carried it around the kitchen for a bit. When I opened the night cage she took it inside with her, and later polished it off quite happily—in addition to her suppertime chick.

MID APRIL

I came home in the evening to find that she had caught another mouse or vole. She was dragging its headless corpse around in her beak, and when she saw me she started making a great triumphalist fuss.

7 MAY

She caught another field mouse; she must have killed it this morning, because I put her out at breakfast time and noticed it when I visited her at about noon. It was perfectly unmarked apart from a broken neck and a neat, surgical incision up one side of its chest. She cawed boastfully while flying round the aviary at top speed, making loud, flashy landings and waving the corpse about. At last she stashed it inside her hutch. She spent more of the day than usual up and about, sitting slitty-eyed on perches watching the nearest greenery. In my mind's eye I could see the thought-bubble: "Here I sit, the keen-eyed hunter—don't come onto *my* patch unless you're tired of living!" In the evening she fished her mouse out again, did a repeat performance of the laps of honour, and finally took it to her shelf to eat.

MID JUNE

One night when I went out to her I found her obviously
excited, but could not work out why. She was hopping
around, squeaking, and "pointing" at something just
outside the aviary. Poking among the long grass in that
direction, I finally discovered a small crashed pipistrelle
bat at the foot of the wire mesh. I found this distress-
ing; it was a delightful creature, and one wing was
hopelessly broken. Mumble was clearly hoping that I
would discover my inner Nero; I rebuked her, shocked,
and—wincingly—put the little thing out of its misery.

◉◉

WHY DIDN'T I feel equally sentimental about the rodents
that occasionally wandered into the aviary, through the
thick grass and weeds that I allowed to grow up both inside
and outside the mesh?

To me, Mumble was a delightful pet; to any field mouse
or bank vole that happened to look up during its last two
seconds of life, she was an unspeakable nightmare—huge
against the sky, lightning-fast, utterly silent, with great
staring eyes, and eight enormous talons reaching wide
enough to envelop its head or grip half-way around its body.
When she hit them like a truck she compressed the whole
power of her foot, leg, and chest muscles into the minute surface area of her claws. Immediately, she would crush the
skull or bite down to break the neck; at least they must have
died almost instantaneously. I suspect that I might have
been tempted to react differently if I had ever seen them

alive in their last moments, but in fact the only evidence of their fate that I ever got to see was after the deed was done (and sometimes the only trace of mouse-icide was a suspiciously dark-coloured pellet the following day).

Every now and then, when I went out to say hello to Mumble after getting home from London on a spring or summer evening, I would find her sitting on her private perch like a feathery Buddha, four-square and solidly planted, breast puffed out, with a preoccupied, upturned face and slitted eyes—and with a small tail dangling from the corner of her half-open beak. On one of these occasions she was having such trouble swallowing this al fresco meal that she actually bounced up and down on her heels while trying to gulp it down. I found myself laughing out loud, and could feel nothing but pleasure that Mumble was tasting at least one of the satisfactions of an owl's natural life.

◎◎

DIARY: 22 JULY

Drama this morning. For some extraordinary reason, a thrush squeezed its way in through the feeding-hole of the aviary, and got a shocking surprise when it discovered who lived there. "Mobbing" was one thing, but mobbing close up and single-handed quite another. Mumble was horrified by all the thrashing about and screaming, and flew from perch to perch to keep away from the wretched thing. Eventually I had to put her in her basket and take her into the kitchen, returning to chase the hysterical thrush out the aviary door

before bringing Mumble out again. Once I had expelled the suicidal fool she behaved with dignity, being too embarrassed to refer again to this unfortunate
incident.

29 JULY

It's a drowsy summer teatime, and I'm enjoying it in a
deck-chair. Some time ago I caught a movement out of
the corner of my eye, and turned to see a large and
characteristically fearless grey squirrel hopping down
the lawn, pausing to grab the occasional snack. It then
ran up the fence beside the oak tree, from there to the
tree trunk, and skittered up to the first branch. Mumble was on her doorstep perch, motionless, and following it with big eyes; she was her usual rotund shape,
not in the tall, thin romaine lettuce format she adopts
when expecting serious trouble. On the fence-top
close to the oak trunk sat Buster the cat. He looked
up longingly at the squirrel, but couldn't convince
himself to try to climb after it—long experience has
probably taught him the futility of following squirrels
up trees.

The three animals held this tableau for moments on
end. It's odd, but there's no question about it: Mumble
still reacts to any other cat that comes into the garden
with a gunfighter's suspicion, but these days she seems
to be almost entirely relaxed about Buster. Professional courtesy, as between lawyers?

◉◉

TO CALL A carnivorous animal "cruel" for playing its part in Nature's constant cycle of eat or be eaten is self-evident nonsense (so far as we know, humans are the only animals capable of conscious cruelty). But that doesn't mean that we can be unmoved by the spectacle of suffering. Personally, I'm grateful that the wildlife cameraman's film is edited to end as soon as the wolves bring down the young caribou—I have no desire to watch the drawn-out horror that follows until the poor creature finally dies, however natural it is.

While I never had any illusions about Mumble's true nature, she did not give me a real demonstration of it until a quiet, sunny Saturday afternoon one May, when a large wood pigeon was eating a newly seeded patch of my lawn. I clapped and shooed it off, repeatedly, but it always came back—until it made the fateful mistake of landing on the mesh roof of the aviary. Mumble had been dozing in her usual shaded private corner next to her hutch, camouflaged by the thick ivy above and on two sides, but by now she was fully alert. As the pigeon settled on her roof she came out of cover in a killing rush, did a mid-air back-flip below it, and struck up through the mesh at it with both feet.

She got most of her talons deep into the pigeon's body, and hung there belly to belly with it, supported by her gently beating wings. Both birds were silent, but Mumble was open-beaked with excitement, and increasingly spattered from feet to face with ruby-red blood that glinted in the sunlight. The mesh in between the two birds prevented a clean kill, and Mumble's locked claws prevented the pigeon from escaping. This stalemate might have lasted all afternoon; I couldn't allow such a medieval death, so I dispatched the pigeon with an air-rifle pellet through the head. Persuading Mumble to let go of it took a long time, but eventually she got tired of hanging on her back in mid-air, and I took the remains into the aviary for her.

She pounced on the corpse avidly and started trying to butcher it, but with little success. City tawnies catch feral pigeons, but this big wood pigeon was a larger kill than a rural tawny would normally make in the wild, with tougher feathers (countrymen will tell you tales of hearing their shotgun pellets rattling off a pigeon's wings at long range). Mumble was not really trained for dealing with a bounty like this. She repeatedly stood on it and tried to pluck a bit of it with her beak, but it kept rolling over and spilling her off. She did eventually pluck and eat about 20 per cent of it, but by that Sunday afternoon she had given it up as a bad job, and I tossed the rest into the bottom of the hedge for one of her less fastidious four-footed colleagues to find.

◎◎

NOWADAYS REAL INTIMACIES were limited to the lazy weekend mornings when we shared the kitchen, preening each other and enjoying each other's company for a couple of hours at a time. I still took great delight in this, and was relieved that at most times of the year it did not take more than a few minutes for us to reforge our mutual closeness, even after five-day intervals when—like a busy couple working different shifts—we really only saw one another when we passed in the doorway morning and evening.

In December 1982 this semi-detached relationship was tested to the full when I spent a month away in Cape Town, courtesy of my generous friends Angus and Patricia. I was ready for it, being unwell after a period of insane overwork following the Falklands War. (As a professional editor, I cannot resist boasting about one of the symptoms—I had begun to leave blood on the typewriter keys.) The month passed in a pleasant haze of late lie-ins, cool wine, and long, rambling conversations in warm sunshine amid the jacaranda trees and bougainvillea, on a hillside above the sparkling blue of the Indian Ocean. I also enjoyed several evenings in a pub that I later discovered is well known to many long-distance travellers—the Brass Bell, in the old railway station on the ocean front at Kalk Bay. Some of these sessions of beer, steaks, and rock-'n'-roll were happily spent with my nephew Graham, who had made the trip the hard way, riding a motorbike all the way from Hampshire to the Cape of Good Hope.

Meanwhile, Mumble was spending the whole month in her aviary, where she was faithfully fed by my kind neighbours—at some risk to their fingers when poking her chicks through the feeding hole (she was too impatient to wait until the mail had actually dropped through the letterbox). I had feared that this long separation might lead to a permanent estrangement, but when I came home she knew me the moment I came in sight, and when I went inside the aviary she jumped to my shoulder at once. I had no trouble persuading her to get into the basket, or to spend the next several hours free in the kitchen while we got to know one another again. She went around the room whooping into all possible holes and re-exploring the nooks and crannies, before settling happily on her perch on top of the larder cupboard.

◎◎

MY WORKING LIFE changed quite radically in the mid-1980s, when a colleague and I started up our own publishing company. At first we produced a military history magazine, and later a growing list of books, from an attic office high above Gerrard Street in London's Chinatown. As anybody who has tried it knows, launching an underfunded infant business and trying to steer it through choppy economic seas is about the scariest and most relentlessly demanding occupation on earth (excepting, of course, those that involve live ammunition). With these added responsibilities my working days grew longer and more intense, punctuated by business trips such as the obligatory, health-wrecking October week at the Frankfurt Book Fair.

Still, Mumble and I continued to make much of one another during our weekend sessions, especially during the summer moulting season. At other times of the year, particularly the winter months when she would have been mating and nesting, she might be aloof, or might even briefly relive her roughest "teenage" behaviour. It was on an evening during one of these latter periods that I fell prey—ridiculously—to a combination of too much red wine and a rather feminine mood of insecurity about our relationship.

On a still, cold, starry night I had fetched Mumble to bring her indoors and was walking up the garden carrying her basket. When I had gone into the aviary she had been stand-offish, but not belligerent. Suddenly, I was seized by a mad impulse to test her true feelings. (Yes, I know—there was no excuse; life had long ago taught me that this sort of thing is always a bad idea.) I stopped, opened the basket, and encouraged her to jump on my shoulder. We stood there together for perhaps ten seconds, while she looked around at what must surely have been a perfect night for an owl. Then she kicked herself into the air and flew up to a branch of the old plum tree about six feet above me.

She seemed calm just sitting there, so, with my heart in my mouth, I walked very slowly away and went indoors, leaving the back door open behind me. During the next very long minute or two I called myself several kinds of fool. Why, in the name of sanity, had I deliberately put myself back in the same situation as on that awful night in London many years ago? What would happen if she heard an irresistible rustling in the grass in the next field—or if another owl called nearby?

Then there was a quiet rush of beating wings, and she flew straight in through the door and onto my shoulder again. All right—it was suppertime on a cold night, and she was no fool; but in my relieved pleasure at feeling her feathers against my cheek again I chose to tell myself that she had acted out of more than simple hunger.

◉◉

AND SO OUR years together in Sussex rolled by and accumulated, and I find that I was seldom making diary entries during this period. Mumble's routines of life as a country lady were well established, and with the exception of keeping a log of her moults and seasonal mood-swings I only felt the need to jot down observations if she did or experienced something (or ate somebody) unusual.

She was a familiar presence in our lane, and my nearest neighbours told me that they liked hearing the reassuring country sound of her occasional calls. Once or twice I noticed local children trying to peer across the garden fences or through the field-hedge to look at her. On these occasions I would tell them to get their parents to ring and arrange a convenient time for me to invite them all into the garden to meet Mumble properly, and to get the "induction lecture" about Tawny Owls. Like me, not all of them had grown up in the country; their first exclamation was always "*Aaah!* Isn't it *lovely*!" and their first question was always "What does it eat?" (They were also unfailingly amazed to hear that Mumble enjoyed taking regular baths.)

When I went into the aviary on these occasions she would fly to my shoulder for reassurance, landing rather hard; the presence of strangers always agitated her, and after a moment she would usually pounce forward to cling, with wings fanning, to the wire mesh closest to them, which usually made them jump. I would use this opportunity to stress that she was very much a one-man bird—and still a wild animal, not a big brown budgerigar. When the kids dispersed I got the impression that there would be a certain amount of boasting when they got to school the next day, but I did not give this as much thought as perhaps I should have done.

10

◉◉

Departure

B Y FEBRUARY 1993 Mumble was approaching her fifteenth birthday; she was not visibly ageing in appearance or vigour, and her behaviour seemed unchanged over several years past. The longest-lived Tawny Owl in captivity reached an astonishing twenty-seven years, and there seemed no reason to doubt that Mumble—safe, sheltered, and well-fed—would get a fair crack at that record.

(I used to tell friends that I had a fantasy about eventually retiring to a house with a tower, in which I would set up my study. In this fantasy, peasants walking home from the fields at night would pass the dark tower with just one lit room at the top, and cross themselves nervously when they saw the silhouette of a bearded figure with an owl on his shoulder— ideally, against a background of flickering green flames. If you are going to be old, you might as well be scary.)

The notebook records that on 5 February that year Mumble showed the expected first signs of the restlessness that I had noted during previous annual mating seasons. During the winter months since October she had behaved towards me in a generally uninvolved way, and weekend caresses had been permitted only after quite lengthy refamiliarization training. She had usually been quite calm, but when I went into the aviary that evening she did a mild demonstration of what I noted in shorthand as "HHS" (hooting and head-shot). She repeated the hooting and the jump to my head when I let her out of the night cage on the morning of the 6th, but when I lifted her down on my arm she didn't do the whistling war-dance that would have added "+WWD" to my note. Instead, she sat quietly in the crook of my elbow and allowed me to nuzzle her head for a bit before she flew up to her perch.

A couple of weeks later, there is a note that when I opened her night cage in the kitchen on the morning of Saturday, 23 February, Mumble flew straight out and landed on my head, but she allowed herself to be lifted down at once, and there was no arm-kicking. After she had gone through her cycle of exploring the room, crapping from the tray-perch, stretching, and having a rudimentary grooming session, I was delighted to discover that she was in the mood to enjoy the familiar weekend morning routine. This included a prolonged and very enjoyable "snogging" session, with her sitting on my lap and holding her face and head up to be thoroughly nuzzled.

Later, while I was eating my fry-up, she was marching around at the far end of the long kitchen table, kicking up unpaid bills like autumn leaves, when she decided that she wanted to be on my shoulder. It would only have taken her a hop and half a wing-beat to cover the yard between us, but instead she chose to walk it—right across my full English breakfast. Crooning softly, she then climbed up my chest, leaving a line of fried-eggy footprints up my bathrobe, before settling down to lean contentedly against my ear. ("Good *grief*, Mumble . . .").

◎◎

DIARY: 25 MARCH 1993

Mumble died last night in the aviary.

☉☉

IT WAS A crisp, starry night. I had gone out to her at about midnight, but she showed no interest in coming indoors, so I fed her in the aviary and left her to it. She had grabbed the chick and carried it around for a while in her beak, bugling defiantly.

When I went out to see her in the morning before leaving for London, I found the aviary door pulled wide open. There was no padlock (what a complacent idiot I was), but it was fastened with a sturdy, stiffly fitting hook-and-hasp that needed positive force with two hands to open it; neither the strongest wind nor any animal could have opened the door like this. Mumble was nowhere to be seen, and I felt an immediate suspicion. I had noticed in the press that some sort of self-appointed animal rights group was publicizing this as a "week of action," and I wondered at once if some pig-ignorant sentimentalist had decided to "free" my owl from her "prison." I had heard nothing during the night; but then, I had slept right through the 1987 hurricane, and my bedroom was at the front of the house. I was completely confident that if anyone but myself had tried to go into the aviary Mumble would have attacked them furiously out of the darkness, and I allowed myself to hope that some local teenager was recovering from a usefully educational fright, while trying to explain away eight deep gashes on his or her scalp.

In the meantime, as badly as I wanted to stay at home, I realized that there was no point in searching for Mumble in the trees of the garden and the nearby fields during daytime.

She would be tucked into the thickest cover she could find, sleeping through the daylight hours; I could go out late that evening, and try to tempt her into calling and showing herself. So I went to work, but I couldn't concentrate, and in the afternoon I left London early. It was only when I got home that I made a really thorough search of the aviary. I had, of course, checked out her enclosed hutch that morning, but I had not thought to comb through the thick, tangled, knee-deep greenery on the ground.

And so it was that I found her; she was lying face down, wings and tail outspread, almost hidden in the middle of a clump of daffodils. There was not a mark on her, and she and the flowers around her were completely undisturbed—which again argued for a human rather than an animal intruder. All the evidence pointed to her having died instantaneously, in mid wing-beat, from a heart attack (to which mature raptors, with their high-protein diet, are always at risk). If any strange human had come inside the aviary she would have flown around in frenzied rage and excitement, and it was plausible that this could have stopped her fast-beating little heart in the blink of an eye. Lord—how I *hoped* she had marked the intruder cruelly . . .

I picked her up and carried her indoors; her head flopped loosely, and when I held her softness to my face I found that I had a clogged throat and stinging eyes.

◎◎

OVER THE NEXT couple of days I wondered what to do with my owl. At first I considered putting her in the freezer and finding a vet qualified to do an autopsy, but there seemed no point—clearly, she had died neither of disease nor of violence. I remembered that many years previously I had wondered idly whether I should have her stuffed and mounted when she died, but now I was revolted by the idea. What would I be left with? A lifeless puppet—a mockery of everything she had been, and a constant reminder of my loss. The thought of simply throwing her body away never entered my mind, and nor could I bear the thought of burying her—what does a bird have to do with the cold, heavy earth?

In the end, I gave Mumble a Cheyenne funeral. I tucked her into the high fork of a leafy tree, with her face towards the hills and sky. The notebook reminds me that, for some reason, at the last moment I felt moved to tuck a few wildflowers around her. I stroked her soft feathers for the last time, pulled the concealing ivy around her, and left her there. When I got home I found myself not just choked up, but sobbing. Before that day I truly don't believe that I had wept aloud in twenty years, and I never have since.

◎◎

A WISE OLD friend of mine once told me that he conceived of our relationship with animals in this way: Mankind has a "vertical soul," capable of touching all levels of existence— from the satisfaction of animal appetites to the intellectual exploration of distant galaxies or the highest flights of artistic creativity, and (since my friend Angus was a believer in a consciousness that survives physical death) that it travelled onwards and upwards thereafter. Animals, he said, have "horizontal souls," in touch as we never can be with every manifestation of life at their own level, feeling and responding to all the tides of which we are unaware—but incapable of upwards movement.

Many ancient folk-mythologies speak of a time "when we all lived in the forest, and people could talk to the animals." Some pockets of humanity—for instance, among the Aboriginal peoples of Australia—seem to retain some sense of what it was like to live at that crossing-point of the vertical with the horizontal axis of consciousness, and to be, at least to some degree, aware of both. Angus believed that it is to our sick cost that the great majority of humanity have lost that "horizontal" awareness entirely, and that even minimal contact with other living creatures and the elemental tides that govern them is beneficial for our mental and emotional health. I did not share his whole belief system, but in that respect I instinctively agreed with him, and my belief was greatly strengthened by my experience of living closely with a wild creature.

Since my childhood I have been fond of both cats and dogs, but before I lived with Mumble I had never had cause to give much thought to my feelings about animals. During the years that we were together her company enriched my life; it saved me from too much self-absorption, and increased my daily pleasure to a degree that I would never have imagined possible. If, from this distance of time, I am to make a rough-and-ready analysis of my feelings about her, I have to start from my feelings about life as a whole (please don't be alarmed—I'm an Englishman, after all, and I shall be brief).

Like, I imagine, the majority of other people of my age

in Britain, I was brought up in the Church of England but drifted away from it in adolescence. I am not a practising or even a believing Christian; nevertheless, there is undeniably a God-shaped hole in my feelings, and I regret that I am unable to believe in a life after death. This regret is not based solely on my honest envy of the evident comfort and strength that believers draw from their faith.

That's part of it, certainly; but in my case, the regret also comes from a vague feeling that in a universe where nothing disappears utterly, but is only transmuted into something else, there ought to be some less wasteful end for something as richly complex as a human personality than simply switching it off and turning its container into leafmould or ashes. It appears to take most of us about seventy years to gain a workable understanding of human life, and to reconcile ourselves to its limits (if we ever do). For that achievement to be thrown away unused, while its container is recycled as fuel for the great engine of physical life, seems a bit spendthrift.

Christians are clearly pretty confused about animals. The hard line is that unlike us, animals have no souls, and so are simply provided as subservient companions—often luckless ones—for us to interact with during our own moral journey towards judgement and an afterlife. (However, some believing Christians have declared that since a Heaven without animals would not be heavenly, it therefore cannot be.) Whatever First Cause you believe in—whether intelligent creation, or chemical accident—logically all sentient life must be the product of the same prime event, and thus all living things must be connected. We travel in company, and a shared journey that divides at the moment of physical death—with one line leading to mere leafmould, and the other going on to some higher destination—seems to me too much like special pleading. It smacks of bureaucratic rules devised by a rather self-important and pettifogging mind, and it offends my sense of the scale of Creation.

Since Mumble and I were evidently both warm-blooded parts of some shared continuum, formed by the same processes and subject to the same basic drives, then it surely follows that either we were both heading directly for the leafmould, or both of us had a further stage of the journey in prospect. I simply cannot feel that I am part of any really fundamental process that excludes her. And if—as seems vastly the more probable—our shared destiny turns out to be oblivion, then I am warmly grateful for the unexpected company of this particularly delightful fellow-traveller.

◉◉

HOW MIGHT A Tawny Owl have felt about travelling with me?

The whole subject of what levels of "consciousness" and "feelings" animals are capable of is a notoriously contentious field of free-for-all argument. Behaviourists, ethologists, and neurobiologists of various schools each have their own orthodoxies (I am tempted to say "ideologies"). Since they cannot agree on a clear definition of terms such as "instinct" and "emotion," there doesn't even seem to be a common conceptual framework for their enquiries. Not having any kind of scientific grounding myself, I can only try to judge my own observations in the light of common sense.

We can never know what it actually feels like to be another type of animal, let alone a bird. There is a constant temptation to project our own emotions onto animals, so I try to stay on guard against this anthropomorphism. I certainly refuse to use the word "love" in any animal context—it's too important for that. At least half of Mumble's walnut-sized brain was a wonderful machine for processing sight and sound, and I can't believe that there was any room in there for abstract thought or for any but the most rudimentary "feelings."

But—and for me this is a huge "but"—she and I evidently enjoyed an individual relationship of some sort, and on her part she gave undeniable signs that it was based not simply on hunger, but on companionship. Tawny Owls are not a gregarious, co-operative species, but they do form long-term pair bonds. There is a mass of recorded observations to confirm that these pairs demonstrate what dry science will only admit are "reinforcement behaviours that reduce stress-hormone levels," but which in ordinary human terms we can only call affectionate pleasure in each other's company and touch.

Mumble's behaviour showed that as she grew into young adulthood she made a clear distinction between me and other humans: if they came near, she hurled herself into the attack to defend our joint territory. Throughout her life she very often chose, unprompted, to seek my physical closeness, and to positively demand my touch, to which she responded with obvious pleasure. If she was startled when we were together she would come to me automatically, staying until she was calm again. She routinely dozed on my shoulder, paying me the greatest compliment that an animal can—that of trust. She very often preened me, as she would have preened her mate or nestlings, and on occasion she even tried to feed me.

Rationalize it however you like; that's an individual relationship—and it's the kind of bond that I have never enjoyed with any other animal, previously or since. I am not interested in analyzing it any further; I am just delighted to remember how very good it felt. Sometimes, all these years afterwards, Mumble still appears in my dreams; and whenever she does, she unfailingly brings a surge of grateful fondness into my mind.

SELECT BIBLIOGRAPHY

BOOKS AND JOURNAL ARTICLES

Birkhead, Tim. *Bird Sense: What It's Like to Be a Bird*. London: Bloomsbury, 2012.

Burton, John A., ed. *Owls of the World: Their Evolution, Structure and Ecology*. Wallingford: Peter Lowe, 1973.

Everett, Michael. *A Natural History of Owls*. London: Hamlyn, 1977.

Hirons, G.J.M. "The effects of territorial behaviour on the stability and dispersion of Tawny owl (*Strix aluco*) populations." *Journal of Zoology* 1.1 (1985): 21–48.

Hosking, Eric, and Flegg, Jim. *Eric Hosking's Owls*. London: Pelham, 1982.

Martin, Graham. *Birds by Night*. London: Poyser, 1990.

Southern, H. N., Vaughan, R., and Muir, R. C. "The Behaviour of Young Tawny Owls after Fledging." *Bird Study* 1.3 (1954): 101–10.

Southern, H. N. "Natural Control of a Population of Tawny Owls (*Strix aluco*)." *Journal of Zoology* 162.2 (1970): 197–285.

Sparks, John, and Soper, Tony. *Owls; Their Natural and Unnatural History*. London: David & Charles, 1970.

Thomas, Paul. "Getting Wise." *Radio Times*. BBC, 1983, 22–28.

Wardhaugh, A. A. *Owls of Britain and Europe*. London: Blandford Press, 1983.

WEBSITES

www.owlpages.com/articles
www.owls.org
www.davidnorman.org.uk
www.javierblasco.arrakis.es
www.raptorfoundation.org.uk/finding.html

ACKNOWLEDGEMENTS

I am grateful to Dick, Avril, and Graham; to Tom Reeves, for photographic help; to Jane Penrose, for her valuable advice; to Christa Hook, for her superb illustrations; and to my agent, Ian Drury of Sheil Land Associates, for his confidence that I could find my way through these unfamiliar woods.